HOW TO DESIGN/BUILD
REMOTE CONTROL
DEVICES

No. 1277
$21.95

HOW TO DESIGN/BUILD REMOTE CONTROL DEVICES

BY IVAN G. STEARNE

TAB TAB BOOKS Inc.
BLUE RIDGE SUMMIT, PA. 17214

FIRST EDITION

NINTH PRINTING

Printed in the United States of America

Reproduction or publication of the content in any manner, without express per-
mission of the publisher, is prohibited. No liability is assumed with respect to
the use of the information herein.

Library of Congress Cataloging in Publication Data

Stearne, Ivan G.
    How to design/build remote control devices.

    Includes index.
    1. Remote control--Amateurs' manuals. I. Title.
TK9965.S76   620'.46   81-9093
ISBN 0-8306-0005-1   AACR2
ISBN 0-8306-1277-7 (pbk.)

Table Of Contents

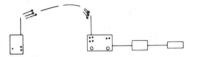

Introduction

When the subject turns to remote control, too many persons tend to cast a suspicious eye in the direction of the man or woman who started the discussion in the first place. Remote control reeks of supremely complicated cams, pulleys, and electronic circuits that only a genius could begin to understand. The purpose of this book is to put this misconception to rest and to explain in easily understood language just what remote control is all about and how the reader can design, build, and use these systems to good advantage. No, the projects are not highly complicated nor are they expensive. Many of them can be built in a few hours and for next to nothing. The jobs they will perform, however, may prove invaluable.

The fact is that all of us are surrounded by remote control devices. We use them everyday. Indeed, we could not get along without them. The caveman didn't, Columbus didn't, and we certainly don't. Basically, remote control is the accomplishment of a manual function from a point removed from the actual area where the work is performed. To do this requires the simulation of many human-mechanical motions. Fortunately, the circuits and the mechanical devices are already with us, having been designed years or thousands of years before our time. Since the means is at hand, all that is required is the knowledge of how to use the many remote control circuits and mechanical devices already present.

This book will provide you with the *understanding* needed to apply remote control to everyday life. After the information contained herein is learned and absorbed, it will become second nature. You will see remote control all around and will immediately recognize ways to apply these principles to other duties. The only other element required will be an active imagination and a concerted effort.

I hope you find this book to be valuable, enjoyable, and—most importantly—meaningful in what you may accomplish in your future endeavors.

Ivan G. Stearne

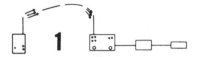

Remote Control

Before discussing the many aspects of remote control and the systems which accomplish the various control initiations and functions, it is necessary to define the term "remote control." In the pure sense, remote control is the initiation and completion of a function or functions normally done at the point where a predetermined *reaction* also takes place. Switching on a radio by twisting a knob on the front panel is direct control. The on-off functions are performed at the point of reaction from these functions.

On the other hand, if the radio switch was left in the on position and an extension cord run to a remote power point, by plugging the wiring into an electrical outlet or removing it from the same outlet, we arrive at a crude form of *remote* control. Here, the radio is switched on or off from an action that takes place at a remote site, remote from the radio receiver which is the central point of action. Remote control, then, is the ability to influence the actions or reactions of a circuit or device from a point which is removed, directly or indirectly, from the unit, proper.

The reaction in this case is the reception of broadcast signals. The reaction is pre-arranged by design of the receiver circuitry. The action is the application of power to the circuit, which is accomplished by tapping a wall outlet at a point that has been removed from the reaction circuitry.

Look around you. You will see that many of the everyday actions most people perform involve remote control in some shape or form. Figure 1-1 shows the basic circuitry of a section of common household wiring. The desired function here is to cause an overhead light to illuminate. Electricity from the house main is

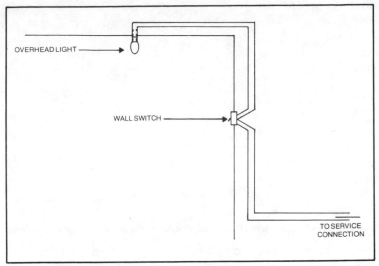

Fig. 1-1. Basic circuit for control of overhead light.

routed to the light through the internal wiring in the wall and ceiling. The flow of current is controlled by the light switch which has been placed at a remote location in regard to the incandescent bulb. This remote location is desirable, because it places the control of the bulb at a point which is convenient to the control initiator—in this case, the household resident.

Getting back to the original definition of remote control, an action must be performed in order to cause the subsequent reaction. Here, the action is the throwing of the switch to the on position, and the reaction is the lighting of the bulb.

Notice that no power was introduced to the system by the initial action. Power was not supplied by the action of throwing the switch. Rather, an already existing power source was tapped and directed through the incandescent bulb. Power—human power—was required to throw the switch. The available electrical power was introduced into the bulb circuit by applying power of another nature to the switch.

The process just mentioned is a very simple example of remote control, but most of the factors involved in this action/reaction sequence hold true for even the most complex forms of remote control. Normally, remote control processes do not introduce power into the system where the reaction occurs. Rather, it routes another, standard power source to the device or devices where reaction is desired. This rule of discrete power systems does not always hold true for all remote control systems if the term remote control is taken in its most literal sense.

An example of remote control where only one power source is in effect is shown in Fig. 1-2. This is one of the most basic power systems known to man and is a very definitive form of remote control. The figure shows the lever/fulcrum system which allows power to be transferred directly into a remote control function. The fulcrum provides an actual transformation of the human force directed at one end in conjunction with the lever. Part of this transfer involved the method and intensity with which the force required to push the lever down is applied to the weight. In this process, the pushing motion of the human body on one end is transformed into a lifting motion on the other.

The fulcrum and lever are a true *machine*. Like all machines, a certain amount of power is required for operation, but you never get quite as much out of a machine as you put into it regarding power alone. While the weight on the end of the lever might prove impossible to lift by hand without the lever and fulcrum, this occurs not because adequate human power is unavailable, but because the power cannot be applied with the force which provides the proper advantage for lifting, using direct lifting methods. More actual power is applied to the pushing end of the fulcrum than is required to lift the weight by the direct method, even though the latter may be physically impossible. In this case, there are only minute differences in applied power and delivered power, but other systems may entail far greater losses.

While the principle of the fulcrum will be explained in more detail later and the reader will gain a full understanding of power versus advantage, this example brings out a very good point: Whenever power is applied to a remote control system, there is a definite factor of loss involved in transferring this power to the

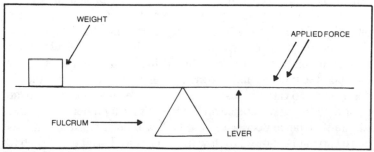

Fig. 1-2. Basic lever.

load. In other words, to apply the most *power* to the weight in this example, the person doing the lifting (or attempted lifting) would have to directly apply the lifting force (his body) to the weight. He would have to grasp it with both hands and pull upward. But it may be impossible for him to *apply* the needed force to clear the weight from the ground. The fulcrum allows him to apply the power needed in such a way as to effect lifting, whereas by the direct method, he could expend three times the power needed for the actual lifting process, because he is unable to direct the force to the correct points of the weight. Another way of saying this is the lifter is unable to gain the proper *advantage* to effect lifting.

Back on the subject of remote control using the lever and fulcrum, lifting the weight is the desired reaction while the action encompasses the human force being applied to the opposite end. The main source of power was not applied directly to the weight but transferred through the machine. This is remote control because a desired reaction was obtained by initiating an action at a point removed from the load.

This differs from the light switch example in that the action not only initiated a desired reaction but also supplied all of the power required for the entire process. No other power system was used. This is not normally the case with most remote control systems discussed in this book, but it serves to provide a better picture of what remote control can and does entail.

While the light switch controlling the on-off reactions of the overhead light is a primitive form of remote control, it is more appropriate to say that this is *remote switching*, a form of remote control. The light may be turned on or off from a remote location, but the control ends with these two functions.

Figure 1-3 shows an example of true control as another component is added to the switching circuit to provide another remote function. Here, a variable resistor is placed in the AC circuit in series with the bulb. As the resistance value is varied, the intensity of the glowing bulb is varied or controlled. Figure 1-4 shows one of these controls installed in the space provided for a normal on-off light switch. The switch and the resistor are incorporated into one unit. To switch the light on or off, the knob is depressed. To vary the resistance factor, the knob is rotated to the right for a lower resistance and to the left for a higher resistance. As the resistance decreases, the bulb will burn more brightly as more current is allowed to flow in the circuit. Turning the knob to the left decreases the brightness by increasing the resistance.

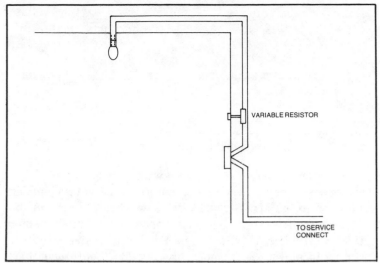

Fig. 1-3. Brightness control is obtained by adding a variable resistor to former light circuit.

Using this device in the light circuit, the homeowner can create several reactions of the bulb from his remote location at the control. The light may be turned on, off or varied in brilliance from very dim to full intensity. This is a better example of true remote control function and circuitry.

Figure 1-5 shows another common example of remote control that is used at homes everywhere. This is a simple sprinkler system with the sprinkler connected to the end of a garden hose which, in turn, is connected to the outdoor spigot. This may seem to be completely different from the electrical example of remote control already discussed, but a closer look will indicate that they are almost identical in function.

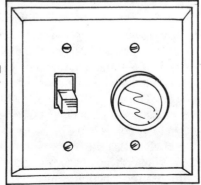

Fig. 1-4. Commercial light control installed in wall switch outlet. Drawing shows complete component.

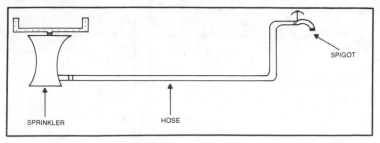

Fig. 1-5. Common sprinkler system for watering lawn.

The power *source* is pressurized water instead of electricity. This source is a separate power system originating at the plumbing connection to the home. The switch now takes the form of the spigot which allows varying amounts of pressurized water to enter the hose. The desired effect is turning the sprinkler arms, *not* watering the lawn. (Remember that this is a quasi-theoretical discussion). The jetting water is merely a by-product of this remote control system.

Again, it is not practical to apply power directly at the sprinkler head, so a control line has been run to a remote location. In this case, the power line is the hose. When it is desired to activate the sprinkler head, the spigot is opened, allowing the pressurized water to travel through the hose (circuit) to the head. Through a series of valves in the head, the flow is directed across a crude flywheel, causing the arms to rotate. The remote function has been initiated and achieved. A sprinkler system of this design is simply a water-powered motor.

Now, what about the true desired effect of distributing water to the lawn. This is a by-product of the water motor, because the power system is a *total-loss* design, regarding the drive element, which is the water itself. Once this fluid reaches the outlet ports on the sprinkler arms, there is still adequate pressure to drive it into the air—geyser fashion—finally becoming deposited on the lawn. *Total-loss* means that once the driving element has accomplished its task, it is removed completely from the system. The water (in this case) is the driving element, and the pressure behind it is the power. The spigot can be directly compared with the variable resistor and switch in Fig. 1-3, because it turns the sprinkler on and off and also controls the intensity of the speed of the rotating arms. It is obvious that the true purpose of this example is not to just turn the sprinkler head but to water the lawn, but we could forego this latter desired reaction by making this a *closed* system.

Figure 1-6 shows how this theoretical example is accomplished. The sprinkler arms are plugged at every point where water might escape. An outlet port is cut into the nonrotating base. Whereas the earlier example was not a true circuit, this one is, because the incoming water is returned to the original system by the return hose. The water can then be repressurized and used as the control element again. The same reaction is obtained using this sytem: The water motor is activated as before. What purpose could this serve? This should be obvious after a little more discussion.

Let's examine the two systems again. The electrical system used an electric switch and resistor to initiate the flow of electricity and to control the flow, while the water or hydraulic system used a valve to initiate and control the flow of water. Both systems are even more similar when considering the fact that electricity in housewiring is also a flow. Electrons within the wiring flow to create electric current. That is why it's called current: It flows. Air pressurization was the source of applied power in the hydraulic system, while the voltage was the power source in the electric system. The higher the water pressure is, the greater the water flow is; the greater the voltage is, the greater the electron flow is (assuming electron quantities are adequate). The resistor and the valve are identical, system for system, because they both control the rate or quantity of flow of the power elements of each circuit.

Now, comparing external power systems and power conversions, we find that to initiate the flows in each system, human control and human power is essential. It takes a power expenditure to flip on a lightswitch and to turn the control knob for brightness, just as it does to twist the wheel of the spigot. These actions released the power or drive elements into the circuits. Human power initiated the release of an alternate power source. The

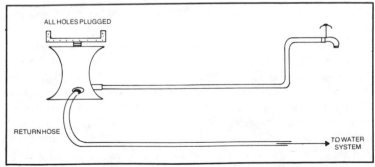

ALL HOLES PLUGGED

RETURN HOSE

TO WATER
SYSTEM

Fig. 1-6. A closed water pressure circuit.

electric current was converted into heat and illumination by the element within the bulb, and the water pressure was converted into mechanical motion of the sprinkler. In the electrical system, we went from human power to electrical power to light power (illumination). In the hydraulic system, we went from human power to hydraulic power to mechanical power. In both cases, the human power was completely removed from the main sources of power and served only as a triggering and control element, but a remote human action, coupled with the entire systems, resulted in initiation of and control of a different power source from the ones activated by the original human power effort. In both examples, the human power system was tapped to allow a segregated power source to be directly engaged and converted into alternate sources of power. Such is the way remote control usually operates.

Getting back to the previous example of a closed sprinkler motor system, the advantage should become readily apparent. Mechanical motion (human) has been ultimately converted into another means of mechanical control and with much more power than the human power required to activate and to control the system. The added power came from the conversion of the tapped hydraulic system. If a spool or capstan were added to the top of the rotating sprinkler element instead of the horizontal arms, a mechanical advantage could be had. Figure 1-7 shows such an arrangement, which might be used to raise a flag on a pole. Here, the flag hoisting rope is attached to the sprinkler spool. When water power is applied, the head and spool rotate, winding up the hoisting rope and pulling the flag into place.

In this example, another conversion has taken place. The motion of turning the spool is lateral, but a downward motion is required to pull the near end of the hoisting rope which, in turn, must be converted to an upward motion of the rope end attached to the flag. Pulleys are used to accomplish this mechanical transformation.

Like the lever and fulcrum, a hoisting system using pulleys enables force to be applied to the correct advantage to accomplish a function or reaction. But the entire system has once again changed. The desired reaction is not to turn the motor but to raise the flag. This remote control system has been expanded, making it much more complex and changing completely the desired reaction of the overall system.

Suppose a system is to be built along these same lines which has as its desired reaction the generation of electric current. This

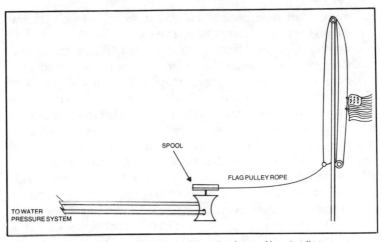

Fig. 1-7. Energy transference and transformation is used to raise flag.

would be a simple process, because the basic part of the system remains the same. Figure 1-8 show how this reaction might be obtained while still preserving the remote control aspects of the water pressure system.

Everything is the same up to the spool at the end of the sprinkler motor. Instead of a hoisting rope, a continuous belt is attached to the track of the spool and through the drive wheel of a belt-driven electric generator. When the generator shaft (connected directly to the drive wheel) turns, the unit generates electricity. Once again, the system is remotely controlled by the operator turning the water spigot to the desired point for proper sprinkler motor speed. As this motor turns, its spool does likewise, pulling the belt in a continuous loop across the generator drivewheel, which also turns. This motion is transferred to the

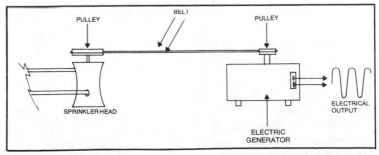

Fig. 1-8. Mechanical energy is changed to electrical energy through system transference.

generator shaft or armature, causing it to spin, and electrical current is generated at its output terminals.

In this example, human effort has controlled the water-pressure system, causing it to power the sprinkler motor that transfers its mechanical energy through the belt to the generator shaft. This power is converted by the generator circuitry into electrical power. The human effort now controls electrical power, instead of mechanical power. The electric power generation process is converted from water pressure power to mechanical power to electrical power, but the original human efforts simply allowed the process to start and to be controlled; it did not *provide* any power to the system which caused the generator to operate.

A similar system could be designed (on miniscule proportions) to allow human power to directly control an electrical generation system. Figure 1-9 shows how this might be accomplished. This is a total drive element loss system which incorporates a small rubber hose to direct pressurized air on the vanes of a miniature windmill with its shaft attached directly to a tiny electrical generator. Human power is used to compress air in the lungs, expelling it through the hose which directs the air on the vanes. This causes the windmill to turn and the generator shaft along with it. When the shaft turns, electricity its produced. Admittedly, only a few microamperes of current would be produced by this system, but it is a good (if not practical) example of a remote control system using a person as the initiator, controller, and main power source. The human is removed from the area of the desired reaction (the generation of electric current). He starts the system by exhaling into the end of the hose. This air pressure is converted to mechanical motion by the turning windmill, which is propelled by the air striking the vanes. This mechanical motion is then converted to electric current.

Returning to the previous discussion on the lever and the fulcrum, it should be noted that if the main power source (the person) were to exhale the compressed air in the lungs directly onto the vanes of the windmill, the results could be laughable. While even more power is being generated by the human, this power would not be applied in such a way as to attain the proper advantage. The hose, then, becomes a machine which allows power to be applied to the advantage needed to drive the system. Less human energy is ultimately expended, because this power is used more *efficiently*. If it took a power output of 1 watt of human effort to power the system, and the reaction or electrical output resulted in a measured power rating of one-half watt, then the entire system is said to have an efficient factor or rating of 50 percent. The loss of the other 50 percent occurred in the effort required to produce the air which did not enter the hose at the human end (that which escaped around the mouthpiece), the air which was expelled from the other end of the hose but did not strike the rotating vanes, the energy expenditure required to start the turning of the generator shaft from a dead stop, and the circuit losses of the generator itself.

POWER VERSUS ENERGY

So far in this discussion of remote control of various types of power systems, the words energy and power have often been used interchangeably. This was done for the sake of simplicity; however, the two are not the same. Power is the *rate* at which energy is transferred or at which work is accomplished. If a person pushes a 2-pound weight a distance of 5 feet, the power required to effect this move would depend on several factors: how fast the weight was to be moved this distance and the resistance factor encountered. This latter factor is very difficult to calculate and will vary for each work load.

Resistance comes in many forms, but the two main resistance factors are friction of the object to be moved with the platform on which it rests and the resistance of the air around the object. The first resistance factor depends on the surface structure of the platform the weight rests on and the surface area of the weight where it comes in contact with the platform. The physical volume

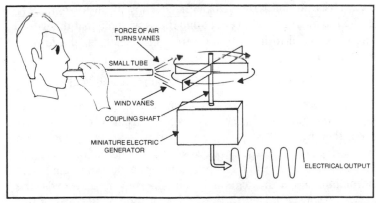

Fig. 1-9. Rubber hose used to direct human air power on vanes of electrical generator.

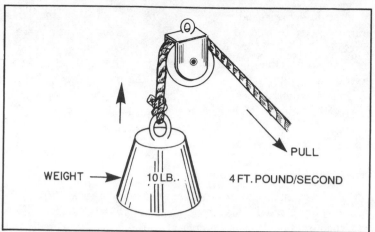

Fig. 1-10. Simple hoist.

of the object will determine its resistance with the air surrounding it. Density of the atmosphere along with the wind direction and velocity will also affect the resistance factor. It is obvious that a 10-pound dictionary could be more easily carried a distance of 5 feet against a 30 mile-per-hour wind than could a piece of paneling measuring 8 feet by 4 feet, even if both objects weighed the same. The physical size and structure of the paneling would cause it to have a much higher wind resistance than the compact dictionary. Surface area determines air resistance. The larger the surface area is, the higher the resistance factor is.

Figure 1-10 shows a simple hoist which, for the sake of this discussion, is designed to lift a 10-pound weight a distance of 2 feet. A basic formula is used to determine how much power must be applied to the rope end to effect this lift. Assuming that the weight was to be lifted to the 2-foot altitude in 5 seconds, the formula would read:

$$\frac{2 \, \text{feet} \times 10 \, \text{pounds}}{5 \, \text{seconds}} = 4 \, \text{foot-pounds per second}$$

This means that the person pulling on the rope would have to supply 4 foot-pounds of force per second for a period of 5 seconds to effect this lift. Foot-pounds per second is a common method for expressing power. Large power systems may be rated in foot-tons per second. If a force of 4 foot-pounds per second were applied to the pulling end of the rope and maintained for five seconds, the weight would be raised to the 2-foot altitude at the end of this period.

This is a practical formula for many remote control calculations, but it is a theoretical formula and will not hold exactly true in actual use. Figure 1-11 shows two examples of the same lifting arrangement used previously. In Fig. 1-11A, the weight is of low surface area. The arrows depict air resistance. Figure 1-11B shows another weight which has a much larger surface area. Notice that the indicated air resistance is much greater for this second weight than with the first, even though they are identical in weight. In practifcal applications, the latter object will require more foot-pounds per second of lifting power than that which is required for the first weight.

If the two surfaces are very smooth, a much smaller amount of movement resistance will be present, and less power will be required to effect the move. The formula used to arrive at the power requirement does not take resistance into account. Figure 1-12 shows the same weight which has had casters or wheels attached to its base. These devices are designed to decrease surface resistance. Less power should then be required to move the weight. But the wheels add a small amount of weight to the object and must be pulled along with it. If the wheels weighed too much, they could actually increase the power requirement, rather

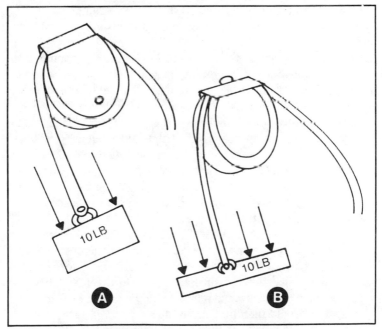

Fig. 1-11. The effects of air resistance on a work load.

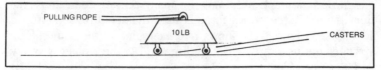

Fig. 1-12. Directional transformation using a hoist.

than decrease it. The formula must be used again to figure the weight of the wheels (or any other object or objects used to decrease friction) and the weight of the object to be moved. It would read:

$$\frac{5 \text{ feet} \times (\text{pounds} + \text{wheel weight})}{10 \text{ seconds}} = \text{foot-pounds per second}$$

In this latter equation, the resistance factor has been eliminated entirely from the calculation, because most of the resistance has been removed. There will still be a resistance factor here, but when proper friction-decreasing devices are used in small applications such as these, this factor can often be ignored.

MACHINES

Reference has been made several times to machines. Most persons think of a machine as a fairly complex mechanical device. The reference to a hose or the lever-and-fulcrum as being machines always raises a few eyebrows. What is a machine? It is *any* mechanical device which is used to transform any mechanical force as to magnitude or direction. The small hose used to blow air from the human lungs across the vanes of the generator windmill is a machine, because it transforms the magnitude or pressure of the air from the lungs. Air is blown into the end of the tube with great pressure. The tube is smaller than the lungs and throat which are supplying the air volume. Therefore, the velocity of the air through the tube or hose must increase to distribute the input air. The hose is also bent for direction of the output air onto the vanes. The hose is a machine because it changes both the magnitude and the direction of the input power. The hose, lying unused on a shelf, is not a machine. But when it is used for the purpose described, it becomes a machine in the true sense.

The rope used to pull the 2-pound weight in the previous figure would probably not be classified as a true machine as presented, because it only transfers mechanical force. The magnitude and direction remain unchanged. However, suppose the rope were bent around a fence post or pole as shown in Fig. 1-13. This system would then be classified as a true machine; not the rope alone, but the rope *and* the post. A transformation of direction

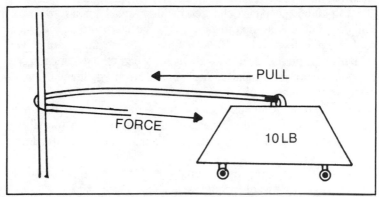

Fig. 1-13. Rope is made into a true machine offering directional transformation by bending around a post.

has certainly occurred in this example. The direction of the force applied to the pulling end of the rope is at a 90 degree angle to the force applied to the weight or work load.

FORCE

Force differs from power only in the fact that the time element is not considered. Force and energy, for this discussion, can be considered to be one and the same. If 10 foot-pounds per second of power are needed to accomplish a lifting, pulling, or pushing process, then the force or energy required is equal to 10 foot-pounds. In other words, 10 foot-pounds of energy must be sustained for 2 seconds to arrive at a power rating of 10 foot-pounds per second.

Taking this into account, if a motor or other such machine was rated to deliver a maximum of 10 foot-pounds per second, it is also rated to produce a maximum energy or force of 10 foot-pounds. Force, then, is measured in foot-pounds. The duration of time this force is applied is called *power*, which is measured in foot-pounds per second.

Now, how much power would be required to lift a weight of 10,000 pounds a distance of 1 foot? We cannot figure this unless we know how much time we are to allow for the lifting process to take place. For instance, if we could allow 10,000 seconds for the lift, the foot-pounds of power would be:

$$\frac{10,000\,\text{pounds} \times 1\,\text{foot}}{10,000\,\text{seconds}} = 1\,\text{foot-pound/second}$$

If 1 foot-pound of force could be applied to this weight continuously for 10,000 seconds, the weight would reach the 1-foot elevation point in 10,000 seconds. This seems like a small amount of power and it is, but look at the power which is necessary to do the job: 1 foot-pound of force times 10,000 seconds, or the amount of power which would be required to lift the same weight the distance in one second, 10,000 foot-pounds per second. The energy level is lower, because the total required to do the job is spread out over a long duration of time. Remember that power is the rate at which work is accomplished. It will take the same amount of energy or total force to move the weight, regardless of the amount of time it will take to get the job done.

Taking a more practical example, to lift a weight of 10 pounds a distance of 3 feet would require an energy output of:

$$3 \text{ feet} \times 10 \text{ pounds} = 30 \text{ foot-pounds}$$

This is true whether the lift takes 1 second or 2 years. How much power is needed would depend upon this time element. If the job were to be accomplished in 10 seconds, the power needed to supply this force would be:

$$\frac{3 \text{ feet} \times \text{ten pounds}}{10 \text{ seconds}} = 3 \text{ foot-pounds per second}$$

Energy or force and power are all interrelated. If a job is to be done in a short period of time, more power will be required than if it is accomplished over a longer period, but a set amount of energy can be required to do the same job regardless of the amount of time required.

THE WATT

A more conventional unit of measuring power is in watts. The watt is an electrical unit which expresses the mechanical term of foot-pounds per second in terms of how much *electrical power* must be generated to produce the former value. The watt, then, is the unit of electrical power required to do work. This work can also be measured in the mechanical term of foot-pounds per second. In an electrical circuit using alternating current, the wattage rating is the voltage times the current in amperes.

Taking this into account, a specific job can be calculated as to its wattage requirement by first figuring the foot-pounds per second rating and then multiplying this figure by 1.356 to arrive at the equivalent in wattage units. For example, a weight of 10 pounds is to be lifted 3 feet in 5 seconds. What electrical power in watts is required to perform this work. The formula would read:

$$\frac{10 \text{ pounds} \times 3 \text{ feet}}{5 \text{ seconds}} = 6 \text{ foot-pounds/second}$$

$$6 \text{ foot-pound/second} \times 1.356 = 8.136 \text{ watts}$$

The wattage rating is used often in remote control application where the work load is to be accomplished through an electrical power system. This rating tells how much power must be supplied by the electrical system to perform the work and allows the person designing the system to choose components accordingly.

LEVERS

While the lever and fulcrum have already been discussed briefly in this chapter, these components are extremely important to remote control applications and they bear a much closer look and study. A lever is a machine. It is used to reduce the amount of applied force required to produce a desired mechanical motion. The use of a lever is called leverage. Leverage means advantage. This is exactly what a lever provides: *mechanical advantage.*

The fulcrum is the point where the lever is supported. Without this component, there can be no leverage. The lever system has two basic points, the input and the output. Figure 1-14 shows these points. The input is where the force is applied while the output is the point where the force is *transferred*. The work is accomplished as the *effort* is applied at the input.

The lever also contains two sections of length which are used to determine the mechanical advantage of the device. Figure 1-15 shows the *power arm*, which is the distance from the fulcrum to the input. This is sometimes called the *force arm*. Also in Fig. 1-15, note the second length, the load arm. This is the distance from the fulcrum to the output.

The mechanical advantage provided by a lever involves simple calculations employing the power arm and the load arm. Friction and air resistance will affect these calculations in practical applications, but in most instances, so minutely that these last two factors can be ignored. The formula for determining the mechanical advantage of a lever is:

$$\frac{\text{power arm}}{\text{load arm}} = \text{advantage}$$

To calculate for any lever, all one needs to do is measure the length of the power arm and divide the length of the load arm into this first measurement. Figure 1-16 shows a lever with a power arm of 12 feet and a load arm of 6 feet. Substituting these figures into the formula, we arrive at:

$$\frac{12}{6} = 2 \text{ (advantage)}$$

The mechanical advantage is two. This means, using the particular lever shown, two times the force will be applied to the load at the

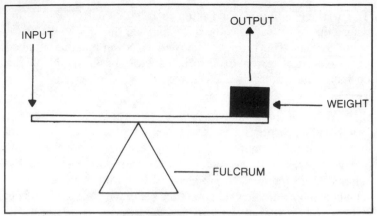

Fig. 1-14. Simple lever with input and output indicated.

output than is applied at the input. If the input force is 4 foot-pounds, the output force will be 8 foot-pounds.

At first glance it would seem that we have gotten twice as much out of the system than was put into it. This is absolutely incorrect. We are actually putting *more* into the system than we are getting out—not much more, but a little more due to the inherent mechanical losses in any mechanical device. We still have to put as many foot-pounds per second of power into the input as we obtain at the output, figuring on a no-loss lever system. To effect the proper lift, the power arm must be moved *twice* the distance, in a leverage system with an advantage factor of two, than we have actually moved the weight or work load.

Several workings of the power formula will bear this out. Figure 1-17 shows a lever with a mechanical advantage of two which is used to lift a 20-pound weight a distance of 1 foot in 5 seconds. Knowing the advantage factor, we also know that the power arm must move twice the distance of the work load, or 2 feet. We also know that the mechanical advantage of two allows us

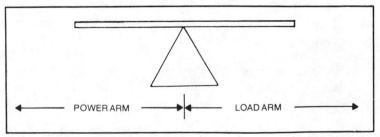

Fig. 1-15. The sections of a simple lever.

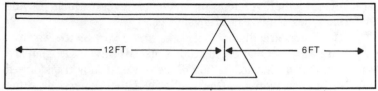
Fig. 1-16. Lever with specified dimensions and work load.

to use half the foot-pound force at the power arm input. To lift a 20-pound weight without a lever would require a force of 20 foot-pounds, but with the leverage advantage factor of two, we can apply only 10 foot-pounds of force at the power arm to lift the 20-pound weight. Substituting these figures into the power formula, we arrive at the following equations:

Load Arm: $\dfrac{20 \text{ pounds} \times 1 \text{ foot}}{5 \text{ seconds}} = 4 \text{ foot-pound/second}$

Power Arm: $\dfrac{10 \text{ pounds} \times 2 \text{ feet}}{5 \text{ seconds}} = 4 \text{ foot-pound/second}$

From these formulas, it can be seen that the same amount of power in foot-pounds per second is required at the input as is obtained at the output, but the applied force is reduced by a factor of one-half in this system. If the lever had an advantage factor of four, then the force reduction would be three-fourths. To find out the amount of force required, simply divide the advantage factor into the weight. The result will be the force required at the power arm input to effect a lift. The power needed is worked through the power formula.

Orders of Levers

The simple lever already discussed was used to transform mechanical force, both in motion and in magnitude; however, there

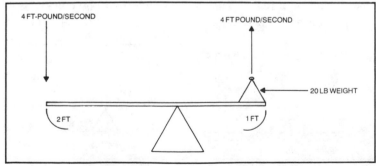
Fig. 1-17. Lever with a mechanical advantage of two.

27

are several other versions or *orders* of levers which provide different types of mechanical advantages.

The lever already discussed is a *first-order* device which is dictated by the fact that the fulcrum is placed *between* the input and the output. A downward motion of the power arm is transformed into an upward motion of the load arm. Other levers place the fulcrum at one end to provide a mechanical advantage in magnitude transformation but not a transformation of direction.

Figure 1-18 shows a *second-order* lever which places the fulcrum at one end while the input is at the other and the *output lies between the two*. The force at the output is greater than the force at the input. Both incur motion of the same direction. As before, the output travels a shorter distance than the input, because when the input is raised, the output section will always lie below it. The power arm section of this system lies between the input and the output, while the load arm is between the output and the fulcrum. The closer the output is to the fulcrum, the longer the power arm is, and, thus, the greater the advantage. The same formula as before will apply here regarding advantage.

Figure 1-19 depicts yet another order of lever. This *third-order* system provides no mechanical advantage whatsoever; in fact, it provides a mechanical *disadvantage*. The force at the output is always *less* than the force at the input. It can be seen that the third-order lever is identical to the second-order example, except the input and output are reversed. While no mechanical lifting advantage is obtained, the machine definition still holds. This device transforms the magnitude of the applied force. It diminishes it. This particular lever is used to increase the *travel* of the load at the expense of greater foot-pounds of force at the input.

One might wonder at this point why the subject of the lever has been discussed so much and so soon in this text. Once a fuller understanding of remote control is gained, it will be quickly seen that the mechanics of remote control far outweigh the problems of

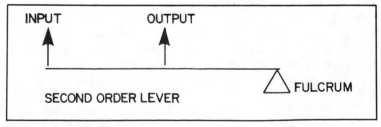

Fig. 1-18. Second-order lever.

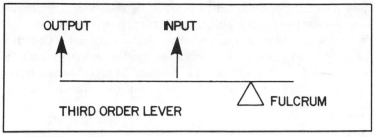

Fig. 1-19. Third-order lever.

providing the proper signals—electronic, hydraulic, or mechanical—to the controlling devices. Satellites orbiting the earth boast tremendously elaborate electronic circuits to allow ground communication, activation, and control, but the mechanical linkages and cams are often far more complicated in design and operation. What the remote control experimenter tries to do in many instances is to duplicate a human motor action through artificial means. The simple act of tying a shoelace would cost many hundreds of thousands of dollars or even millions of dollars if this were to be performed by a mechanical equivalent. In forming the schematics for various mechanical "hands," it is necessary to at least have an elementary understanding of force, power, and leverage.

If you will look around you, you will certainly notice that many, everyday household devices use levers or, more likely, a series of levers. A lightswitch may well be a simple lever. The most obvious lever would probably be a child's seesaw. As a child, you may have made a crude seesaw from a long board and a large rock (to serve as the fulcrum). No doubt, you also had the experience of trying to enjoy this pasttime with a playmate who was much larger than you. To compensate for the weight difference, you probably moved the power arm farther away from the rock. You sat on this end and were able to ride up and down for hours, even though an imbalance seemed to exist. By moving the power arm toward the lighter person, less force was required to lift the weight on the other end. With the rock fulcrum at the exact center of the board, this lever offered a mechanical advantage of one, which is not mechanical advantage at all, but by increasing the length of the power arm, the lighter person gained a definite advantage.

Chances are that you also noticed (if you were the light one) that when you were at maximum altitude at the top of the lift, you were much higher in the air than was your seesawing partner. If your seesaw lever had an advantage factor of two, you had to travel twice the distance of your partner. Isn't it amazing that the mathematical formulas discussed in this chapter are subjects which we all have dealt with by instinct in the past? Many other things can be done with levers and the lever principle. These will be discussed in later chapters.

PULLEYS

Another type of machine which has been discussed only briefly so far is the pulley, or rope and pulley. Figure 1-20 shows this machine consisting of a small wheel and cover, which provides a low-friction track for the tope to ride on. The rope and pulley form a machine, because, in the figure shown, a downward pull on the rope results in an upward lift of the weight on the other end. Our machine has transformed the direction of the applied mechanical force. This particular example does not provide any mechanical lifting advantage. If the weight is 10 pounds, then 10 pounds of force will be required on the other end of the rope to effect the lift.

The pulley principle does not need a standard pulley to be applied. Figure 1-21 shows how this principle is adapted using the branch of a tree as the substitute pulley to still effect a directional transformation. The friction factor here will probably be much higher, however, and additional pulling force will be required.

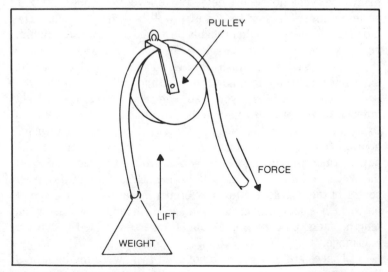

Fig. 1-20. Rope and pulley.

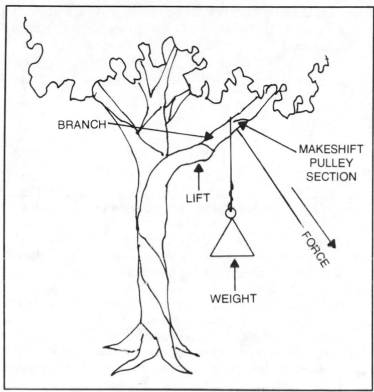

Fig. 1-21. Adaptation of rope and pulley using a tree branch.

The rope and pulley can be used not only to obtain a directional transformation but a lifting one as well. To understand this, it is first necessary to understand some of the physics behind energy transfer. Without going into extreme detail, one law of physics states that the force in a rope is equal at all points. This force is often referred to as tension. Figure 1-22 shows a rope pulling a 100 pound weight. Assuming a perfect situation of no friction between the weight and its platform and negating any air resistance, it will take a force of 100 pounds to move the weight. The meters in the rope represent strain gauges that measure tension. All gauges within the rope will show identical readings, because the tension or force within the rope is the same at all points.

Figure 1-23 shows how this principle can be used to mechanical advantage when applied to a more complex rope/pulley system. Here, two pulleys are used. One end of the rope is attached to the top pulley. The cord is then passed through the

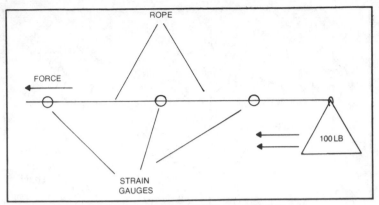

Fig. 1-22. Single length of rope used to pull 100-pound weight.

bottom pulley, up through the top pulley, then down and up again, repeating the process. When the rope is passed through the top pulley for the second time, it is located in a configuration which allows access to the power source. Assuming the weight totals 400 pounds, this same figure in foot-pounds of energy is required to effect a lift, but only 100 pounds of force is required at the pulling end. The pulley system has offered an advantage factor of four. This occurs because the 400-pound weight is supported and pulled by four strands of the rope. Remember, energy in a length of rope is equal at all points. The 100 pounds of applied pressure at the pulling end or power input of the rope is transformed. The force is mechanically magnified by the rope and pulley.

As before, you don't get something for nothing. The input end of the rope must be pulled a distance of four times the required lifting distance of the weight. If the required lift is to an altitude of 2 feet, the input end of the rope must be pulled a distance of 8 feet. By substituting these figures into the leverage formula of an earlier discussion, the result will show that the same number of foot-pounds per second were required at both the input and the output ends.

It must be continually realized that these calculations are meant to apply only to perfect leverage systems with no resistance. In practical use, there is, of course, a resistance factor. So in this example, it will actually take a small percentage of extra force to lift the weight. For most systems, this percentage is so small that it is not normally made a part of the calculations.

We can see from the discussion on levers and pulleys that remote control offers certain distinct advantages aside from being able to perform work at a distant location. Remote control is sometimes necessary to take advantage of the transformation aspects many of these devices require. It may not seem like remote control when you lift a weight with a pulley or even more a large rock with a lever, but in each instance, a job is being performed at a distant point from the immediate area where the work occurs.

MORE ON ENERGY

Before moving on to the more practical applications of remote control, a brief discussion of energy law is in order. *Energy can neither be created nor destroyed.* This is a hard and fast rule which, as far as we know, applies to the entire universe—again, as far as we know. An electrical generator does not actually generate or create energy. It simply transfers energy from one system to another. A generator works by physical motion: A shaft turns through a series of copper conductor windings. An electrical field is formed by the magnetic energy of the rotor material cutting through the conductor windings. Without the rotary movement of the shaft, the mechanical energy could not be transferred. It takes as much energy to turn the rotor as the device generates or transfers. In practical application, it takes *more* energy to turn the shaft than is

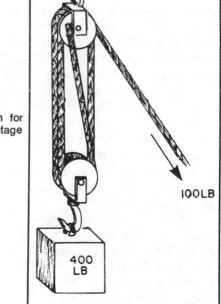

Fig. 1-23. Two-pulley system for increased mechanical advantage through force magnification.

100LB

400 LB

produced in the electrical output. This holds true in every practical application. You must put more energy into a generating system or into a work load than you get out of it. This, again, is due to friction, ohmic (heat) losses and many other factors. The difference between the input power and the output power is the efficiency of the machine.

But what happens to the power which is not transferred? It went in the input but didn't come through to the output. Was it destroyed? No. Energy cannot be destroyed, only transferred. And that's exactly what happened. The energy was transferred out of one system and into another. But there was only one other system: the electrical generator. Right? Wrong. Friction losses transform the input energy into heat. This heat energy is transferred to the atmosphere where it might be absorbed or stored by the earth (another energy transfer). It might even be stored in wood pulp or plants, only to be later released as the air cools. The energy is transferred back into the atmosphere. Sunlight is stored in the plants and trees it nourishes. When these trees are cut down and used for firewood, the energy is transformed into heat through combustion, the heat is transferred to the atmosphere, and the cycle starts all over again in this and a million other ways. Everything in the universe has stored energy which, given the right set of conditions, can be transformed or transferred to another system. This goes on forever.

With all of this energy around us, it may seem strange to hear of the energy shortage. This is not a true statement taken in the most literal sense. There is a shortage of material which we normally use to provide us with a transfer of energy from only a few of the many millions of possible sources. There is about the same amount of energy on the face of the earth today as there was a million years ago. The amount would be exactly the same save for losses through radiation into space and radiation from space to earth. It would be safe to say that there is theoretically the exact amount of energy in the universe as there was eons ago. The amount has not been increased nor has it decreased—again, theoretically speaking.

What about the electric energy used to power a motor. Isn't that destroyed? No, again a transfer has taken place. The electrical energy is transformed into mechanical energy represented by the spinning shaft. Some of the electrical energy is passed off as heat to the atmosphere due to ohmic losses (resistance) in the conductors. Friction from the air creates heat as the shaft spins. Friction discuss some of the ways these principles could possibly be put to use in remote control applications. Throughout this part of the discussion, constantly compare these practical applications with the theory already learned. You will see how theory is used in almost every step to arrive at a complete remote control system.

Figure 1-24 shows a scheme for opening a door by remote control. This would be a practical scheme because an air compression automatic return is used to close the door when the opening power is removed from the mechanical circuit. A rope and pulley arrangement is the method whereby the opening is accomplished. The power source is not specified. This would normally take the form of a motor which will be discussed in a later chapter, but for now, let's assume that it is any mechanical or electrical device which has the power capacity to effect the opening. But how much power is needed? Let's assume that the door weighs 20 pounds. It is mounted on hinges which are designed to reduce mechanical friction, so the friction factor will be ignored. The automatic door return will offer minor resistance to opening the door, so that can also be eliminated from the formula for the time being. The door is actually a form of lever when arranged in this manner. The fulcrum is the hinged portion. In this case, the far edge of the door is the input, because this is where the force is applied. The center of the

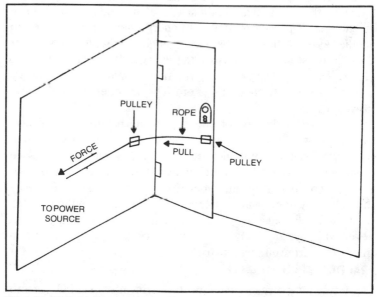

Fig. 1-24. Remote door opening system using rope and pulley.

door is the output. True, no weight is being applied externally to this lever, but the door itself, making up the power arm and load arm, acts as the load. While it is impractical to figure the advantage factor of the door because of the difficulty in determining the weight of the power arm and the load arm, we still know that a greater mechanical advantage is obtained by applying the input power to the outside edge. The door then is a lever of the second order in this configuration. If the power source (the rope) was attached to the center of the door, then this would be a lever of the third order. From the earlier discussion, we know that a third-order lever provides a mechanical disadvantage, because the input force must always be greater than the output force.

Because the door weighs 20 pounds, we will assume that a force of 20 foot-pounds must be applied to effect the opening. The second mechanical machine now comes into play in the form of the rope and the pulley. In this arrangement, no transformation of magnitude is provided because the rope is looped through the pulley only once, but a definite change of direction is provided. The door is to be opened laterally while the force is applied vertically to the door swing. How much power is needed to open the door? We still don't know at this point, because we do not know how far the outside edge of the door is to travel and in what amount of time. The outside edge was chosen for the power connection point for between the shaft and the bearings also produces heat. The spinning shaft passes on until the energy here is transferred to another system. If you think hard enough, you will soon see that there is no stopping or destroying energy. We cannot create it either. Atomic power is simply the transformation of the energy stored in an atom to another system. The law of energy holds true in every case.

Because of the law of energy, no mechanical or electrical system designed can offer an increase in power above the sum of what was originally put into it. Nor can any man-made device produce an energy output which is exactly equivalent to the input energy. To do this would require a perfect transfer of energy through a system which offered no resistance and no losses whatsoever. Because this process hasn't been invented yet (and never will), we must always assume a certain power loss when transferring from one system to another.

PRACTICAL APPLICATION

Having discussed some of the principles behind remote control, power, energy and leverage, it is now appropriate to

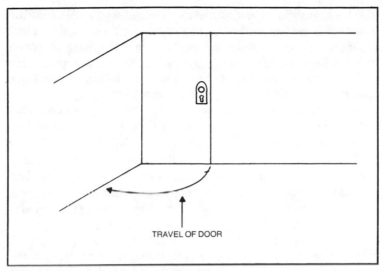

TRAVEL OF DOOR

Fig. 1-25. Travel of outside edge of door in arc.

another reason. The outside edge must travel farther than any connection point between it and the hinges. This gives us a bit more flexibility in opening time if a motor is to be used as a power source. A faster turning motor can be used, whereas at a point closer to the hinges, the motor would have to turn very slowly to effect a safe opening. For practical purposes, a full opening time of 5 seconds should be adequate and safe. If the power source would handle the load, the door could be opened in 1 second or less, but this could be very hazardous to occupants. Remote control usually dictates slow-moving mechanical devices.

Figure 1-25 shows how the travel of the outside edge of the door is measured. It is not a straight line but an arc. For this example, the total travel of the outside door edge will be 4 feet, so 20 pounds of load must be moved 4 feet in 5 seconds. The formula is then:

$$\frac{20 \text{ pounds} \times 4 \text{ feet}}{5 \text{ seconds}} = 16 \text{ foot-pound/second}$$

The power required to do the work is 16 foot-pounds for 5 seconds. This is the power rating of the energy source. If this were a motor, it would have to be able to supply at least this much power. To do this would require a horsepower rating of 0.0288. Horsepower is the normal way in which the power capability of a motor is rated, rather than in foot-pounds per second. The chapter on motors will explain how to make this conversion.

If the available motor turned too fast to open the door in a safe 5 seconds, another pulley arrangement could be used to good advantage. Figure 1-26 shows how the rope is looped through two other pulleys similar to an earlier example in this chapter. This pulley system now has an advantage of four which means that the rope must be pulled at the new power input end a distance of four times farther than before. If the motor speed was previously four times too fast, opening the door in a little more than 1 second, it will now take the desired 5 seconds to effect a full opening.

The door will close automatically when the power source is removed by means of the pneumatic device at the top edge. The project shown here is used in one form or other for automatic- and remote-controlled opening of all types of doors and other similar devices.

Let's refer to the earlier problem of the available motor turning four times too fast to cause a 5-second opening. If the single pulling system were used, the motor would exceed its power rating of 16 foot-pounds per second. The formula used to arrive at the power used 5 seconds as the completion time. If this time is sped up by a factor of four times, the door would be opened in 1.25 seconds. A motor designed to deliver 16 foot-pounds per second of power would be overtaxed. It would require four times the power rating to do this job within safe operating limits. The formula is:

$$\frac{20 \text{ pounds} \times 4 \text{ feet}}{1.25 \text{ seconds}} = 64 \text{ foot-pounds/second}$$

Using the pulley system would slow the opening of the door to the required 5 seconds, but the rope would have to travel an additional length. The total pull would be 16 feet. The formula would read:

$$\frac{20 \text{ pounds}}{4 \text{ (advantage factor)}} = 5 \text{ pounds}$$

Then

$$\frac{5 \text{ pounds} \times 16 \text{ feet}}{5 \text{ seconds}} = 16 \text{ foot-pound/second}$$

The same power requirement as in the first example is obtained. The advantage the pulley system provided allowed the motor with the higher turn rate to spread out its power over a longer period of time, effecting the accomplishment of a job that it, alone, was unable to do without exceeding its power ratings. This is what mechanical advantage is all about. It allows the remote control enthusiast to have automatic, mechanical control of the power band of various devices, electrical, mechanical, and hydraulic.

In practical application, the complex pulley system would probably not be used. The chapter on motors and rotary motion will detail other leverage devices which can transform motion and force in a less complex manner; however, this pulley system would work where, without the system, the job could not be accomplished with the components stated.

Figure 1-27 shows the use of a simple lever to turn a toggle switch to the on position. In this case, the switch lever must be pushed upward. This can be accomplished with a downward motion of a *solenoid*, which is an electrically actuated device with an arm that pulls inward when electricity is applied to its input. A small lever is attached beside the switch and could be made from spring steel. The fulcrum could be a small wedge made from wood, plastic, or metal and epoxyed to the front panel.

This tiny machine transforms motion or direction. The lip of the control arm of the solenoid is hooked over the input edge of the lever at the end of the power arm. The load arm makes contact with the toggle switch lever. When the solenoid is activated, the control arm applies force to the lever which transfers this power to the switch, raising it to the "on" position. In practical application, this system would probably attach the lever to a slip ring surrounding a bolt which is passed through the panel face. The lever would be soldered or welded to this ring in order to be held in place. The

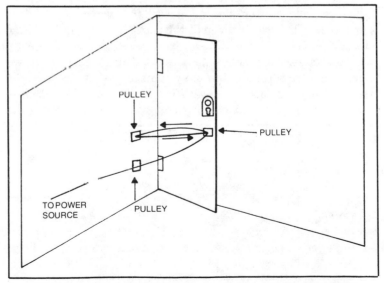

Fig. 1-26. Close-up of rope and pulley arrangement for door opening system.

fulcrum would then be the mounting bolt. This is illustrated in Fig. 1-28.

Solenoids are often not adjustable as to the length of the motion of the control arm. When activated, the arm will travel a set distance. The distance might be too great for a certain application, so the length of the power arm of the lever would be increased. For the sake of discussion, let's say the arm travels a distance of 3 inches, but the switch must only travel a distance of 1 inch. A lever system with the power arm three times farther from the fulcrum could be set up, so a 3-inch travel at the input would result in a 1-inch travel at the output, as shown in Fig. 1-28. There is a mechanical advantage here of three, which means that a large switch might be easily controlled by a small solenoid, since the control arm need supply only one-third of the required force needed without leverage to raise the switch. The switch, itself, is a fulcrum of sorts with its input being the control arm and the terminal point in the switch case serving as the fulcrum.

SUMMARY

If you have absorbed the information presented in this chapter, you should be able to see machines all around you. Levers, pulleys, and combinations of these two devices have been used hundreds of times in almost every home.

Remote control involves the accomplishment of a job at a point which is removed from the immediate area where the work is performed. This usually involves the transference of energy from one system to another. Often, many transfers take place (in fractions of a second) before the work is done. We know that we never get more out of a system or systems than what was originally put into them; in fact, we always get a bit less. The amount of output from a system divided by the power input determines the

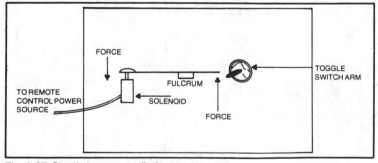

Fig. 1-27. Simple lever as applied to toggle switch.

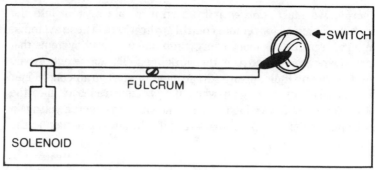

Fig. 1-28. Modified lever does away with separate fulcrum.

efficiency of the system. The higher the efficiency factor, the better the energy transference from input to output.

We have repeatedly stated the law of energy transfers. We have also explained how power can be transferred through mechanical levers, pulleys, and other devices to accomplish jobs which require *force magnification*. The power required to generate this force was the same at the input as it was at the output; only the applied force had been transformed as to magnitude. To transform any action or force to a larger degree requires the increase of other factors, such as time, distance of travel, etc. No power input can be amplified in any way without transferring power from another system.

The mechanics of recreating human actions and movements is the most taxing problem in remote control application. Even the simplest of human actions may create severe design problems when trying to make a device which will perform the same function or functions mechanically. The control of the working device by remote means is simple in comparison to the building of the device itself.

Your knowledge of levers, pulleys and the laws of force, energy, power, and transference will be used over and over again in almost every remote control application. So far, we have only touched on a few of the mechanical motions needed to effect true remote control. Fortunately, even the most complex remote control systems use variations of the lever and pulley (and a few other machines to be dealt with in later chapters) and, often, combine many of these devices to arrive at an overall system of mechanical leverage and control.

Last in our summary is the repetition of the law of energy. Energy can neither be created or destroyed. When we need

41

energy, we must *transfer* it from an available system into the systems required for remote control applications. These systems, in turn, transfer the input energy into other energy systems that make up or are derived from the work itself. The same energy you use to turn on a radio, open a garage door, or start an air conditioner may be used by you again when it is transferred back into the original system from which you obtained it. It may serve you again in a split second or in 50 years, but it will always remain in the universe.

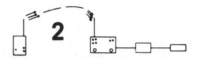

Electromechanical Devices

Remote control requires the reproduction of human actions by mechanical, electrical, or electronic means in many systems. These actions are often difficult to simulate, but there are certain devices which have been standardized and fall quite readily into play with remote control systems. While many mechanical devices may have to be fabricated to take care of the special situations which are not standardized but unique, many situations will require the addition of readily available devices which may be purchased and used as is or modified slightly to arrive at an operational remote control system.

ELECTRICAL SYSTEMS

Figure 2-1 shows an example of remote control similar to what was discussed in the first chapter. Here, it is desired to control an overhead light but, this time, from a very great distance of several hundred feet. It has been mentioned several times that losses occur in any system. Generally speaking, the longer the power path is in the system, the greater the system losses are. While the example in Chapter 1 involved a fairly short path between the power source, the switch and the reaction point (the light bulb), Figure 2-1 involves a much greater distance and thus much greater system losses.

Since this is basically an electrical system, the losses will be incurred as *ohmic resistance* in the power wiring. Ohmic resistance is a value, measured in ohms, which describes the abilities of a wire or other type of conductor to pass the electrical current flow.

While electrical wires are considered to be conductors of electricity, they also resist the flow to a certain extent. This resistance increases as the electrical current causes heat, which is radiated all along the path. This heat is a transference from the electrical system, so less electrical current actually reaches the load. It will take considerably more input power to effect the desired reaction at the output end of the system. The amount of resistance, or *ohms value*, will be determined by many factors, including the composition of the conductor (copper, aluminum, steel, iron, silver, etc.), the conductor diameter, conductor length, and the amount of switches and other devices in the line. A copper wire resistance table is provided in the appendix to help with this factor.

CONDUCTANCE FACTOR

Conductance is the reciprocal or opposite of resistance. If a length of copper wire has a resistance of 10 ohms, the conductance factor is 1/10. Conductance is measured by a unit called the siemens.

The siemens (S) is rarely used to describe a length of conductor. The ohm value is used more or less universally, and the siemens value can be quickly figured from the former unit. A conductor will always pass the flow of current more than it tends to pass the flow. The higher the ohms value, the more the resistance factor. The higher the siemens value, the higher the conductance.

PRINCIPLES OF VOLTAGE, CURRENT, AND RESISTANCE

What effect does resistance have in electrical circuit. This will depend upon many values, including ohms, voltage, and current. We already know that the ohm is the basic unit of electrical resistance, but what about these other two factors? Perhaps these can best be explained by relating them to their mechanical counterparts.

Voltage is the electrical unit of force—electromotive force. Electromotive force is measured in units called *volts*. Electrical current flow involves the movement of electrons in a conductor. The better conductors generally have more closely spaced electrons that are easier to move by electromotive force. Figure 2-2 compares voltage and current flow to baseballs being packed end-on-end in a hollow tube. These baseballs represent the electrons in a conductor. When the ball at one end is struck by a baseball bat, representing the force or voltage, the balls strike each other, driving out through the opposite or output end. The force at the input has produced a like force at the output. This is how electrical current flows in a circuit. Resistance losses can be compared to the stationary baseballs in the tube. It will take an individual consumption of energy to get each one moving, so the total amount of force was used up in getting the process to work. The longer the hollow tube and the more baseballs contained therein, the more resistance there is when compared to a shorter tube with less baseballs. In the former situation, a higher striking force will be required to produce the same amount of force which could have been accomplished at the output of the shorter tube with a less forceful input strike.

Figure 2-3 now compares electrical current flow and voltage to the first-order lever described in Chapter 1. The input or voltage is applied at one point and is transferred to another, the output. The bar or platform represents the conductor and current flow. The fulcrum cannot easily be explained in electrical terms at this point, but future discussions will show that there are ways to provide an advantage factor with electrical circuits. For now, it will suffice to say that a simple electrical conductor with an applied voltage at one end and a subsequent output can be thought of as a type of lever with a mechanical advantage of unity. In other words, no advantage or disadvantage.

OHM'S LAW

Just as there were formulas for determining input versus output force, mechanical advantage, and so on when using the basic

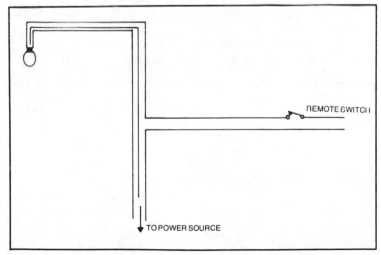

Fig. 2-1. Hard-wired remote activation circuit with remote control point a great distance from work load.

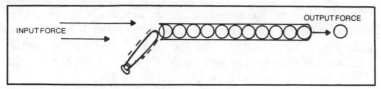

Fig. 2-2. Representation of current flow in electric circuit.

lever, there is also a basic formula which has many derivatives used for determining the effect resistance has on a circuit. This is called Ohm's Law and reads:

$$E = IR$$

You will use this equation over and over again whenever electrical systems are to be used for remote control applications. Each letter represents a different electrical value. The letter "E" stands for voltage, "I" for current, and "R" for resistance. In plain language, Ohm's law states that voltage "E" is equal to the circuit current "I" times circuit resistance "R." This is the basic formula from which all electrical equations stem.

As a working example, assume that the current in an electrical circuit is measured at 2 amperes. The *ampere* is the basic unit of current flow. The measured circuit resistance is 10 ohms. What is the circuit voltage? Simply multiply 2 times 10 and you find the voltage measurement is 20 volts. The formula would read:

$$2 \text{ amperes} \times 10 \text{ ohms} = 20 \text{ volts}$$

By doing some rearranging of the formula, which is stated to figure for voltage when the current and resistance are known, we can figure for any of the remaining two values, resistance, or current, when any two of the three values are known. For example, in the equation, if we knew the voltage equaled 20 and the

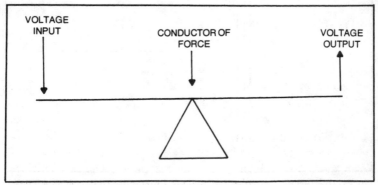

Fig. 2-3. Voltage and current as compared with a lever.

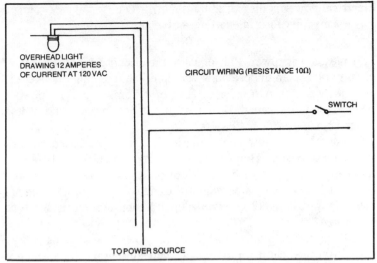

OVERHEAD LIGHT
DRAWING 12 AMPERES
OF CURRENT AT 120 VAC

CIRCUIT WIRING (RESISTANCE 10Ω)

SWITCH

TO POWER SOURCE

Fig. 2-4. Circuit parameters of remote control wiring from distant point.

resistance value was 10 ohms, we could solve for the current by using the modified Ohm's Law formula of I = E/R, or current equals the voltage divided by the resistance. In this case, 20 volts divided by 10 ohms equals 2 amperes. Likewise, if voltage and current are known, the formula of R = E/I or resistance equals the voltage divided by the current applies. Twenty volts divided by 2 amperes equals 10 ohms.

PRACTICAL APPLICATION

Referring to Fig. 2-4, we see a similar light activation circuit to the one previously described, but the voltage and resistance values are known. We know that the required voltage at the light is 120 volts and the resistance is 10 ohms between the power source and the load (light). This is the resistance of the light bulb only, not the entire circuit. What will be the current value through the bulb? Going to Ohm's Law, we know that current is equal to the voltage divided by the resistance, so the formula reads:

Current (I) = 120 volts/10 ohms or current = 12 amperes

At this point, things tend to get a bit confusing. The resistance value is confined to the light bulb only, not to any resistance in the wiring from the power source. If the light bulb did not have resistance, it would not use *any* power; no transfer would occur. The rate at which energy is used is called *power* or the *power factor*.

47

Electrical power is measured by the *watt*. Ohm's law comes into play in a modified form for wattage and reads:

$$P = IE$$

This means that power (P) is equal to the circuit current times the circuit voltage. In the previous example, we found that delivered voltage was 120 volts at a current of 12 amperes. The power transferred (often called consumed) in the bulb alone is 120 volts times 12 amperes, or 1440 watts.

Going back to the example in Fig. 2-4, we know that to deliver the required power, a force of 120 volts at a resistance of 10 ohms requires 12 amperes of current, but let's ask another question: How much input force was required to deliver the output force of 120 volts at the light? This will depend upon total resistance or circuit *resistance*.

To figure circuit resistance, we must add the 10 ohms of the work load, the light bulb, to the resistance of the line. Remember that this is a circuit: Two wires run to the bulb, one delivering current to the load and the other to allow it to flow back to the source, completing the circuit. Let's say the line resistance equals 20 ohms. Before going to the Ohm's law equation to figure total power consumption, it should be pointed out that, in an electrical circuit, current is the same at any point. If 12 amperes are required at the output, then 12 amperes of current must be delivered at the input. We know that circuit resistance is 20 ohms. We also know that circuit current is 12 amperes. What is the total or circuit power needed to provide 1440 watts of power to the bulb? Ohm's law says that power is equal to voltage divided by the current, but we don't know the input voltage or force applied to the circuit. We only know what must be delivered to the load. Unlike current, voltage can vary throughout a circuit, due to power consumption. Remember, resistance losses result in the transfer of force (or voltage, in this case) to another system. Ohm's law can be modified for the power formula as well and while $P = I^2R$ where only circuit resistance and current are known. In this case, the formula would read:

$$P = (12)^2 \times 20, \text{ or } P = 144, \text{ or } P = 2880 \text{ watts}$$

This 2880 watts is double that which was actually delivered to the load (1440 watts). An input of twice the power delivered at the output was required, because the resistance factor was doubled. While a transfer of 1440 watts was made to the heat and incandescent light system of the bulb, a like amount was dissipated as heat losses from the electrical line. This is a terribly inefficient

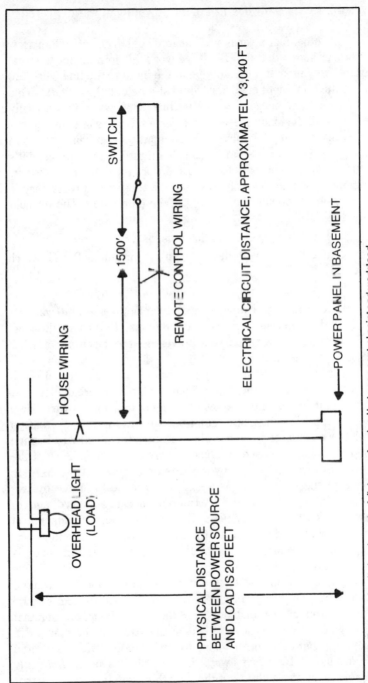

Fig. 2-5. Remote circuit showing actual distances involved between control point and workload.

system, because only one-half of the power input was delivered to the load. The rest was wasted.

One other factor can now come into light. The basic Ohm's law for power states that $P = IE$. If we work this formula for the total circuit, we find that if the voltage at the input to the line were 120 volts, the total input of power would still be only 1440 watts. This cannot be true, because we know that 2880 watts was actually delivered to the input. Because the current is the same throughout the entire circuit, the voltage at the input must be different from voltage present at the work load. To figure input voltage, let's return to the basic $E = IR$ formula and work it for the *entire* circuit. Current is 12 amperes as before, but the total circuit resistance is now 20 ohms, double the reading of the bulb alone. The formula would read:

$E = 12$ amperes times 20 ohms, or $E = 240$ volts

This can be checked with the basic power formula of $P = IE$ which would read:

$P = 12$ amperes times 240 volts, or $P = 2880$ watts

All formulas agree now that to deliver a power of 1440 watts to the load through total resistance of 20 ohms, 10 through the load and 10 through the wiring, a total power input to the circuit of twice that amount is required.

This vast waste on electrical power would be required to remotely switch on the light discussed from a great distance. Actually, it would take roughly 3000 feet of total circuit wire (1500 feet to the bulb and 1500 feet back to the source) to create a resistance factor of 10 ohms. This assumes the standard 14 gauge copper wiring was used. We could increase the efficiency of the system by increasing the size or diameter of the wire used and thus decrease the resistance, or we could increase the force by two times at the light bulb (if it were rated to operate at this voltage) to halve the current requirement. Either way, the same amount of power would be delivered to the load with a decrease in power output over the previous example. Neither of these methods is usually very practical.

Figure 2-5 shows what can happen in a possible remote control situation when a light or other electrical device is to be actuated from a great distance from the home. It can be seen that the power source is as physically near the load as in the example in Chapter 1, but the control wiring for the remote switch is located the discussed 1500 feet away. The circuit has been *electrically lengthened* by the addition of the switch conductors. The current

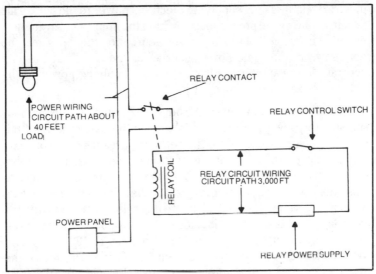

Fig. 2-6. Relay control from remote point using separate power circuit for relay coil.

must now flow from the source, through one control line to the switch, through the switch, back through another control line to the load, and finally back to the power source again. If some way could be found to efficiently control a switch within the home in a mechanical manner by another system at the remote control point, a great savings in power might be had.

RELAYS

Such a way is readily available at almost every electrical outlet and hobby store in the world. It is called the *relay*, and the name is well chosen, because it relays one switching action to another point. Pictured in Fig. 2-6, the electric relay is activated by its own, separate electrical circuit that requires very little power. The input power in this circuit is transferred to a mechanical system, causing switching contacts mounted on the relay to close. These contacts take the place of the light switch on the wall of the home in the example of Chapter 1. Control wiring is then run from the relay coil (its electrical system) to the remote position.

The advantage may not be immediately recognizable. You still have to run the 3000 feet of electrical wiring if you want to establish a control point 1500 feet from the home switching point. The advantage is that the relay will require a power source far smaller than that required by the light bulb to effect operation. The relay electrical circuit is not supplying electrical power to the bulb. This is being done by the power source within the home and only a short distance from the load. The power source for the relay may be derived from the same power source for the bulb, but this is a separate leg or circuit leg from the one supplying power to the bulb. This example is shown in Fig. 2-7.

Here, the relay is powered from the 120-volt AC line that also supplies the power for the load. Whereas the load requires 1440 watts of power for operating, the relay requires less than one 100 watts in most instances. For this discussion, let's assume the relay draws exactly 1 ampere of electrical current for its contacts to close the main power circuit to operate the light. All other factors will stay the same. The control line to the switch that now activates the relay exhibits a total resistance of 10 ohms. The force required at the relay can usually be anywhere from 100 to 130 volts, with most devices designed to operate from the AC line. This means the power consumed at 110 volts, for example, would be 101 watts. (Ohm's law of $P = IE$, or $P = 110$ volts $\times$ 1 ampere, or $P = 110$ watts.) We now have to figure the resistance of the relay circuit alone, which would be $R = \frac{E}{I}$ or resistance equals voltage divided by current, or 110 ohms. Figuring now for the entire circuit, we add 10 ohms (line resistance) to the 110-ohm relay coil resistance and arrive at a total resistance o f 120 ohms in the relay and control circuit. Power is equal to the current squared times the resistance, or 1 ampere (1^2 still equals 1) times 120 ohms. The total power consumption in the line is 120 watts. Now, checking the voltage input to the remote control system, we find that $E = P/I$, or $E = 120$ wtts divided by 1 ampere. Input voltage is 120 volts. We are well within the range of the electrical force provided by the home power system for the input and can still apply an acceptable amount of voltage to the relay.

Using this example, we have remotely controlled a high-power circuit by using a relay with a separate, low-power electrical circuit. Much more efficiency has been gained by using this method. Before, it required a total of 2880 watts at the input to deliver one-half that amount to the load. Using the same system with a relay, total power consumption for operation and control is 1440 watts for the bulb, plus 120 watts for the relay, for a total power consumption of 1560 watts. This is a savings of 1320 watts, or over 45 percent. Since 1560 watts was put into the system and 1440 watts was delivered to the load, the overall efficiency of the total system is a little better than 92 percent. With the previous method, the efficiency was only 50 percent—quite a savings.

The relay and auxiliary devices find much usage in remote control applications where various electrical circuits must be switched on and off. Relays are available in thousands of different configurations, each designed to perform a specific switching function or several functions. The relay electrical system also varies tremendously, and many types and values of electrical power can be used to effect the mechanical opening and closing actions.

Relay Parts and Design

As has already been stated, a relay is a switching device that mechanically closes and opens its switch contacts by converting electrical power into mechanical power or motion. The heart of the relay is its coil, which is shown in Fig. 2-8. When electrical current flows through the windings, an electric magnet or electromagnet is formed. The electrical system is transferred to a magnetic system. This magnetic power can either pull or push something. The voltage and current required to operate the relay will depend upon the weight of what is to be pulled or pushed and the number of turns of wire in the coil.

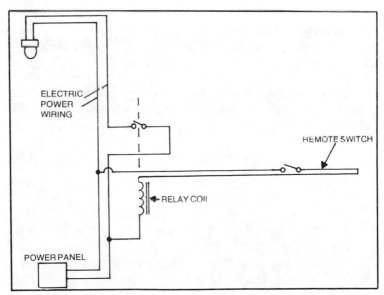

Fig. 2-7. Relay control using AC line circuit for relay coil supply.

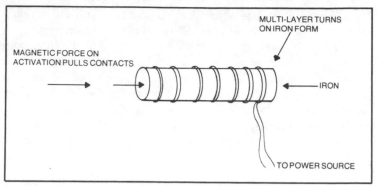

Fig. 2-8. Relay coil arrangement.

The part of the relay which performs the switching function is called the *switching contact*. This part is shown in Fig. 2-9. This can consist of only two, separate pieces of metal, one which moves when the relay is activated, and the other remaining stationary in most cases. The surface area of these contacts determines how much current they can handle. Their insulation from other components in the relay and with the circuit to be switched determines how much electromotive force or voltage can be handled. Contacts with very large surface areas handle larger amounts of current than those with small surface areas. The larger

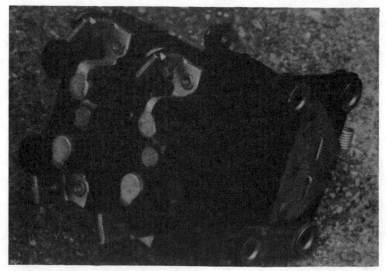

Fig. 2-9. Picture of relay switch contacts.

54

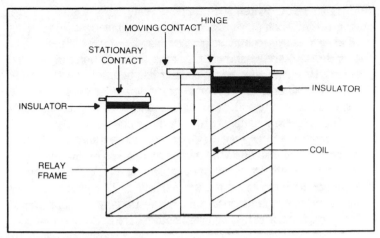

Fig. 2-10. Normally open relay.

contacts weigh more and generally require larger mounting platforms, which increase the power consumption of the electromagnet to effect proper switching.

Figure 2-10 shows how the moving contact is, in this case, pulled by the electromagnet. The base of the moving contact is hinged to the relay frame so that movement may be had when magnetic force is applied by the coil and power system. These platforms or bases are often spring loaded, so when power is removed from the coil, the platform and its contact automatically

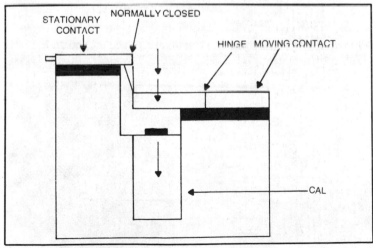

Fig. 2-11. Normally closed relay.

return to the unactivated position. The relay shown remains in the open position when there is no power input. This is called a relay with *normally open* contacts, often abbreviated *N.O.* Figure 2-11 shows the same relay and contact base but with normally closed contacts. Here, the stationary contact has been placed above the movable one where, before, the nonmoving contact was below the one controlled by the coil. Nothing has been changed other than to move the stationary contact to the overhead position.

The relay shown has only one contact arrangement which allows for the control or switching of only one leg or one wire in an electronic circuit. Other relay contacts may offer many more switching functions. The one pictured in Fig. 2-12 provides many separate switching functions from one, basic relay. Here, some of the contacts are normally open while others are normally closed. A relay of this type may turn on several pieces of equipment while, in the same motion, it turns off several others. Multicontact relays often find their way into complex remote control systems when higher amounts of power are handled.

Figure 2-13 shows schematic representations of various types of switching arrangements using relays. Other than the electrical triggering, relay switches are almost identical to mechanical switches and are rated identically as to contact arrangement and current and voltage handling capabilities.

Latching Relays

In the former example of relay control of the lightbulb, it was necessary to constantly supply power to the relay in order to complete the circuit to the load. While a small (by comparison) amount of power was needed to actuate the relay and to keep it closed, this is not always desirable. A standard light switch

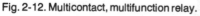
Fig. 2-12. Multicontact, multifunction relay.

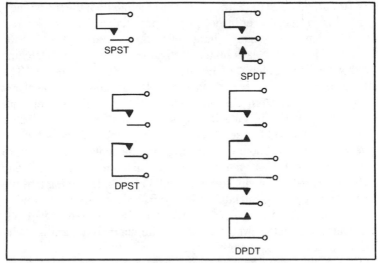

SPST

SPDT

DPST

DPDT

Fig. 2-13. Various switching arrangements provided by relays.

mounted in the wall of your home does not require constant trigger action to maintain circuit completion. The human-mechanical action of throwing the switch lever is momentary. When the switch is pushed to the on position, it stays on until another human-mechanical action turns it off. This is a more efficient manner of control because far less power is used in the control process.

Fortunately, there is a type of relay that duplicates this procedure and still uses electrical power (instead of human-mechanical power) to cause the switching. This is called a *latching relay* and is shown in Fig. 2-14. This device requires actuation current for only a split second. Once the contacts are closed, a mechanical latch snaps to the movable base of the contact and keeps it closed. Power may then be removed, and the switch still remains closed. To open the switch, another pulse of electrical current is passed through the relay coil by once again closing the control switch at the remote location. In this manner, the latching relay exactly duplicates the functioning of a standard light switch by maintaining or breaking the electrical circuit with each, momentary input of power to the switching circuit.

Using the same example as before but replacing the standard relay with a latching variety, we can save even more power. Assuming the latching relay also draws 1 ampere of current at 120 volts input to the control line, it will still take the same amount of energy to close the relay, but look at the time factor. The basic unit of power is the watt. The unit of power for 1 hour is the equivalent of 10 watt hours. This is the electrical equivalent of the mechanical measurement of power in foot-pounds per second.

The previous nonlatching relay circuit required 120 watt-hours of power for each hour of activation; however, the latching relay requires an input of 120 watts for only a split second. The time after contact closure does not involve the consumption of any power (at least from the electrical system), so the savings are monumental. It will take the same amount of power to close a latching relay and to keep it closed for one day as it does for one year. Comparing 1 hour of operation with the nonlatching circuit to the same amount of operation with the latching relay will bring about a watt-hour savings of over 36,000 times, assuming that the relay can be closed and latched in a typical one-tenth of a second or less. This would figure out to 120 watts at one-tenth watt per second or 1/36,000 watt per hour. The control system is many thousands of times more efficient.

For switching functions similar to the example sited here, latching relays are almost universally used. Control with the latter device is more reliable, because the control circuit requires only momentary power input. Because of the decreased *duty factor* of the control line (not carrying current continuously during operation), smaller conductors can often be used which would be severely underrated to handle this amount of power for a continuous period.

Fig. 2-14. Latching relay.

INDEPENDENT CONTROL POWER SOURCES

The switching examples discussed so far with the electric relays have all used the 120 volt AC line, which was also the power source for the work load. This is neither absolutely necessary nor desirable in many remote control applications, and the various types of relays available allow for much versatility in this area. Small latching relays are available that will operate from low-voltage sources of from about 1.5 to 12 volts. These often require minimal amounts of current for switching the contacts. Some operate from as little as 3 or 4 milliamperes (a milliampere is one-thousandth of an ampere). Control lines which carry current at low-voltage potentials are often much easier to install, and these low voltages do not present the safety hazards that are always of concern when using house current. Small relays are ideal for many applications, but they are limited as to the amount of current they will handle at their very small, lightweight contacts. In situations where high-current loads are to be handled, it might be necessary to use a more complex system of relays in the control circuit to close the main power circuit.

Figure 2-15 shows an expansion of the previously discussed remote control system that ultimately closes the circuit of the light bulb but does this in a multisystem manner. In this example, the

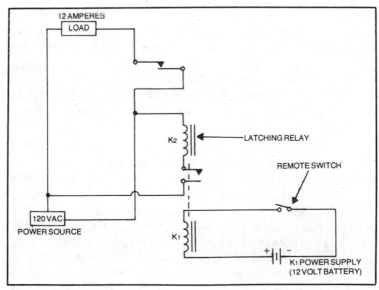

Fig. 2-15. Latching relay controlled by driver relay circuit.

small, low-voltage, low-current control relay does not have adequate contacts to directly close the main power system circuitry that is drawing 12 amperes. The contacts are rated to handle only 2 or 3 amperes and would quickly heat and burn if used to close this circuit, but the original latching relay can be actuated by a current of 1 ampere at 110 or 120 volts. The first control relay, then, is used to actuate the second latching relay which, in turn, closes the circuit on the main power system.

Relay K1 is not a latching relay. It is of the type which requires constant current to remain closed, but it is not necessary or even desirable to keep this in the closed position for more than a second or so, because the power it switches to the latching relay need only be present for a short time to cause the latter to revert to the latched position. All relays are located near the light bulb (within the home). In this example, the first relay obtains its power from a 12-volt battery. The power used by this device is only applied for a moment, so the battery should provide many many hours of switching operation. The control line is run to the remote control location, and the switch at this end simply controls the 12-volt relay line with which it is in series. When the switch is thrown momentarily, the 12-volt relay closes its contacts. This action switches on the latching relay power supply, which is still obtained from the house current. The latching relay closes and locks the main power system circuitry to the light bulb. By now, the power has been removed from the 12-volt relay and thus from the latching relay, but this latter device remains closed due to its mechanical latching ability. The work load of 1440 watts has been switched on by a small relay which can only directly control power systems of a few hundred watts. Since the 120-volt line to the latching relay is now much shorter after being replaced by the 12-volt line to the first relay, another savings of control power has been obtained, making the system even more efficient.

This example brings out a good point. When small relays or other devices are desirable as first control mechanisms in high-power systems, use them to control the electrical systems of large latching relays that are rated to handle the full current load of the main system. In some cases it may be necessary to add a third or even a fourth relay to gradually build from a low-current handling capacity to a medium-power system to a high-power system and, finally, to a ultra-power load. There is no real disadvantage to using this *stepping* method in most applications. The time required to finally close the main circuit will be multiplied, but relays close so rapidly that this rarely becomes a factor. For instance, if 10 relays are used in steps to open or close a system, and each takes a maximum of one-tenth of a second to function, then all 10 relays will be actuated in a total of 1 second. Most relays take much less than one-tenth of a second to close, so this is an extreme example.

Figure 2-16 shows how one relay with several switching contacts can be used to control many switching arrangements in one latching throw. Generally speaking, the higher the number of contacts is, the more current the relay coil will draw to close. This is due to the added weight of the contacts and their movable base. This arrangement would be desirous if it were necessary to, for example, turn on or off several work loads at one time with each work load being powered by a separate source. If some of the contacts were normally open while others were normally closed, one latching position of the relay would turn some remote devices off while other contacts were simultaneously turning on yet other devices. This is advantageous in many situations, but how would you control these devices independently? You could use three separate relays with three discrete control lines and switches, but this gets rather involved. There are easier ways to accomplish these tasks.

STEPPING SWITCHES

A stepping switch is a type of relay. It is controlled by current flow through a coil as with the standard relay and it performs in a latching function, but the switching contacts are arranged in a

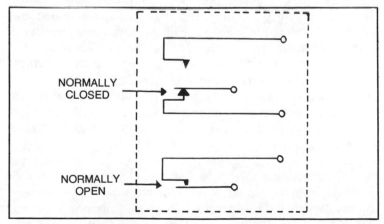

Fig. 2-16. Multicontact relay switching. One set of contacts is normally open, and the other is normally closed.

rotary fashion, so that a different switching function is performed each time the relay is activated. Figure 2-17 shows a typical stepping relay that contains many different switching positions. When power is fed to the coil for a moment, the first contact is closed. This could directly power a work load or serve as a function switch for another relay. When power is again applied to the stepping relay coil, the second contact is closed. Depending on the switching arrangement, the first contact may be opened again or may still remain closed. The progression continues until the switch makes one complete revolution and winds up again at the first contact.

The stepping relay offers the selective control of electrical devices from a remote location. If contact 3 controls a light bulb, then three pulses of power to the stepping relay coil will activate the power circuit to the bulb. All other contacts can be chosen for various other devices to be controlled.

Another problem may be encountered at this point. What if you want to turn on a device which is switched by contact 3? This would mean that you would have to provide three current pulses from the control line to the stepping relay, but you would also turn on the devices connected to contacts 1 and 2, if only momentarily. This is not desirable from efficiency and control standpoints. It is time to introduce another electromechanical device to the stepping circuit to have even better control.

TIMING SWITCHES

A timing switch is a crude clock that closes or opens a contact after a set period of time. The time element is often controlled by a mechanical adjustment. Shown in Fig. 1-18, the timer works on electricity which starts its internal clock. If the device is set for a 10-second *delay,* this means that the contacts on its switch or switches will be actuated (opened or closed) 10 seconds after it begins receiving power. If the power is interrupted during this time period, most timers or *delay switches* will automatically reset and will require another input of power for the full 10 seconds in order to be actuated.

By installing timers in each circuit switched by the stepping relay, immediate control is gained over all elements connected to the rotary contacts. In the previous example of closing contact number 3 without triggering operation of the devices controlled by contacts 1 and 2, the time-delay switches allow this to occur, because the stepping relay must close the contact of the device to

Fig. 2-17. Stepping relay.

be switched and keep it closed for at least 10 seconds for actual operation to begin. To control the device connected to contact 3, three quick pulses of current are applied at the stepping relay input. With each pulse, the relay advances to contact 1, 2, and finally 3, but because the pulses were of less than a 10-second duration, the first two devices were not supplied power, because the time-delay relay in series with their power circuits never closed their switches. The timers were actuated for a second or so, but when the power was removed as the stepping relay switched to another contact, they simply reset their timing elements, waiting for that pulse of 10 seconds or more. When the third pulse was triggered, it would be held for a full 10 seconds or more by the control operator. This would allow the timer to go completely through its 10-second cycle, close its contacts and allow the current path to the device under control to be completed. Figure 2-18 depicts this process.

As is the case with other types of switches and relays, the stepping relay can be obtained with normally open or normally closed contacts or a combination of both. The same applies to the time-delay relay or timing switch which may contain one or several sets of switching contacts. Current ratings also vary trememdously with the size, design, and intended use of each device.

The introduction of relays to an electrical system is an easy method of obtaining remote control capabilities; however, each requires a separate and special electrical installation. What if you wanted to turn a light on and off using the present switch installed in the wall without having to go to the time, bother and expense of installing relays? You would need a device which simulated the human finger and hand actions required to manually throw the light switch control arm.

SOLENOIDS

The function of a relay is to close contacts on a switch that is part of and mounted to the relay frame. While a mechanical action does take place in the switching, this action is not directly introduced as motion to the main power system of the work load. A device which is very similar to a relay but which does not perform the closing action of a self-contained set of switching contacts is called a *solenoid*. The solenoid is an electromechanical device which transforms electrical power into mechanical power or motion. Figure 2-19 shows a typical solenoid which consists of a multiturn winding of electrical wire on a hollow form. In the center of this wound form is the armature, which is made from permanent magnet material or soft iron. When current is passed through the coil turns, it becomes energized and produces a magnetic field. This pulls the armature into the middle of the coil. A spring is often used inside the hollow tube to push the armature back out again when the power is removed and the coil deenergize. Whereas the relay converted electrical power into mechanical power, specifically to close the switching contacts, the solenoid converts electrical power into mechanical motion which can be used for a multitude of purposes.

The travel of the armature is short and extremely fast, so the armature could be attached to a mechanical device to provide a small degree of device movement. In our case, the armature could be attached in some way to the electric light switch, so that its

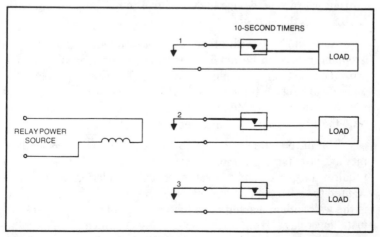

Fig. 2-18. Using timers with stepping relay for practical control of several circuit loads.

Fig. 2-19. Solenoid construction.

inward motion when the coil is energized could control the position of the switch arm.

Figure 2-20 shows how this might be accomplished using a small length of heavy twine connected to the switch arm and then to the end of the armature. When mounted above the lightswitch, energizing the solenoid coil would pull the armature inward and upward. The switch arm would be pulled with it to the normal on position. When power was removed from the solenoid, the internal spring would push the armature back to its original position. If the spring were powerful enough, the switch arm would then be returned to the off position.

Another high-ranking application of the solenoid is in the remote control of door locking. While special *solenoid locks* are available on the commercial market, one can easily be made with a conventional 120-volt AC solenoid designed for general purposes. While the solenoid lock tends to be rather expensive, a surplus solenoid can be obtained for a few dollars and installed in less than an hour. Figure 2-21 shows how this might be accomplished.

The solenoid is mounted at the edge of the doorframe with its armature shaft extending across the door edge into a small "U" clamp. When the coil is energized, the armature is drawn into the coil tube and the door may be opened. Upon removing the power to the solenoid coil, the internal spring pushes the armature back into the locked position. The power switch within the solenoid power line is installed at the remote control point desired. Solenoids are sometimes rated in foot-pounds per second but this is not often the case. These are not momentary devices. Power is required constantly for the armature to be held in the closed position, but the work load is often momentary as to its duration. In the example of a solenoid controlling a standard light switch, once the switch has been thrown to the up or on position, most of the work load is completed. The only strain on the armature is in holding its inward position against the outward push of the internal return spring. For this reason, even small solenoids may be used to move large loads if only for a very short period of time. This may overload the device, but only for a split second. As long as the arms pulls or pushes the mechanical switch or whatever is being controlled without a lot of straining (as can be evidenced by a loud hum), the device should be adequate for the job. The high current drain through the coil occurs the moment resistance to the inward pull is met. Once the pull has been effected on the light switch, it latches mechanically. No further energy is needed to keep it in this position. The device has only been overloaded for a short time, and no damage to the solenoid coil should result.

INDUSTRIAL TIMERS

Similar in some ways to the time-delay relays previously discussed, the industrial timer is usually a standard clock mechanism that is set like any other clock. The device may be

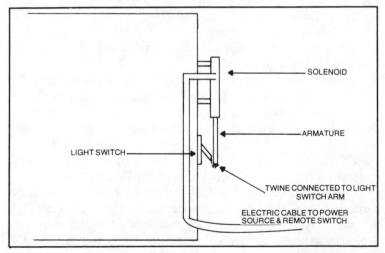

Fig. 2-20. Use of solenoid attached directly to light switch for remote control of power.

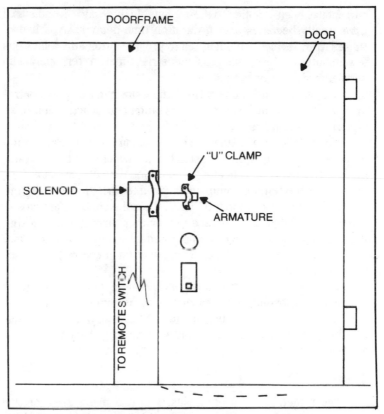

DOORFRAME

DOOR

"U" CLAMP

SOLENOID

ARMATURE

TO REMOTE SWITCH

Fig. 2-21. Solenoid-controlled door lock.

programmed to open or close its internal switching contacts at a preset time or times.

Figure 2-22 shows a timer used for many automation purposes. This device does not provide remote control in the strictest sense. Once it is programmed, it will automatically perform the switching functions without any outside control. This, then, is *automation*, which differs from remote control in that the control responses are programmed into the system before the reaction is to be obtained. The programming is done at the device itself, and once set, needs no further human actions to perform its assigned task.

Industrial timers may be used in remote control systems, however, when the timers are initially triggered by external means. Some timers delete the clock face and can be programmed to effect the switching processes an hour or more after electrical current has been supplied to their inputs. In this way, the industrial timer really becomes a long-duration time-delay relay. Most of these latter devices are designed to provide automatic delays of a few minutes or less, whereas industrial timers often operate in segments of hours or even days.

A good example of use of the remote timer in a remote control system would be in the remote activation and deactivation of an electric stove. Figure 2-23 shows how the stove might be activated from a remote location using a larger, latching relay. The activation process also engages the industrial timer which will be programmed to switch its contacts after a set period time of, for example, 1 hour. After the remote command is given to turn the stove on, the timer will allow operation for 1 hour. Then the contacts are closed and the stove latching relay receives another current pulse causing it to disengage. The stove has operated for a period of 1 hour and then been automatically shut down. All of this occurred with only one remote command.

The use of automation devices in remote control systems is desirable and absolutely necessary for many applications where it is not practical to continuously issue a series of remote commands. For this reason, remote control and automation go hand in hand and are often confused.

SUMMARY

The remote control enthusiast will find many uses for the various types of electromechanical devices on today's market. Each of these transforms power from an electrical system into a

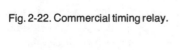
Fig. 2-22. Commercial timing relay.

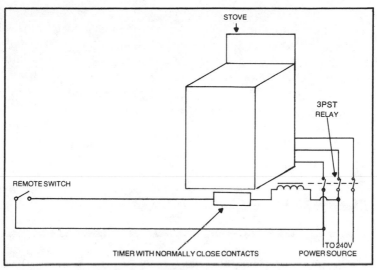

STOVE

3PST
RELAY

REMOTE SWITCH

TIMER WITH NORMALLY CLOSE CONTACTS

TO 240V
POWER SOURCE

Fig. 2-23 Remote controlled activation of electric stove with timing shut-off circuit.

mechanical or motion system. Relays, timers, and time-delay relays all use metal contacts and perform specific switching functions, and solenoids may be used for mechanical movement to control other devices. Each electromechanical device must be chosen with adequate ratings to control the desired system or systems. Some might need to be modified or completely rebuilt in order to conform exactly to a specialized application. For proper operation, the relay contacts must be kept smooth and clean. This assures an efficient, low-loss transfer of electrical current. Burnishing of the relay contacts is the normal maintenance procedure. This involves rubbing them with an abrasive to remove carbon and to keep the contacts even. A burnishing tool is available for this purpose.

Many of the electromechanical devices mentioned in this chapter are available from government surplus markets in a wide range of power systems designed to handle varying amounts of current. When purchased commercially, they might be rather expensive, especially for the high-current models. As surplus, though, they are very cheap.

Perhaps, the best known electromechanical device, the electric motor, was purposely not discussed in this chapter. Its types, range of operations, and many other criteria place it in a category by itself. The next chapter is devoted fully to this device and to its many modes and applications.

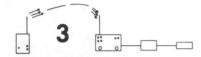

Electric Motors

The levers, pulleys, and other mechanical devices discussed so far have all been useful in transforming mechanical energy and providing leverage and advantage. All of these systems had an input for the application of force and a resultant output. None of these, however, generated its own force. Each only transformed the input energy. On the other hand, one device did generate force by transferring power from an electrical system into mechanical motion which could be used for mechanical control purposes. This device was the *solenoid*.

While solenoids are often used for generating motion, the motion is limited and not continuous or long lasting. When a constant source of mechanical power is desired for a remote control system, the electric motor is usually the universal choice. Like the solenoid, the electric motor consists of coils made from windings of electrical conductor. The motor continues to parallel the solenoid because it contains an armature which is within the coil turns. At this point, much of the resemblance ceases, because, when energized by electrical current, the motion which is set up in the armature is rotary motion as opposed to the back-and-forth, linear motion of the solenoid.

Selecting an electric motor to perform a mechanical function is not difficult when all parameters of the system it is to control are known. It is fairly simple, once this criteria is obtained, to figure out the foot-pounds per second required and then to convert this value into *horsepower*, a term by which most motors are rated.

There are many different types of motors. They come in a range of horsepower sizes to perform various functions. Some are

70

designed to be operated by standard 120-volt AC house current. Others require the input of direct current. Still others will operate from both AC and DC.

Another motor rating is in rpm. This is an abbreviation for revolutions per minute and describes the number of full rotations the armature makes when the coil is fully activated within 1 minute. Most remote control applications around the house will require very low rpm at the force input to the mechanical system. Most motors operate at very fast speeds, so it may be necessary to fabricate a speed reduction system to convert the high rpm into speeds that are more readily used in the remote control process. These systems can be designed using much of the knowledge which has already been obtained from the first two chapters of this book.

ROTARY MOTION

The electric motor transforms electrical energy into mechanical energy in the form of rotary motion. This motion may be in a clockwise or counterclockwise direction, depending upon the internal construction. Some motors are reversible; they can be switched from clockwise motion to counterclockwise motion by reversing the polarity of the applied DC current to their inputs.

This latter trait is very useful in remote control applications. Figure 3-1 shows how a low rpm might be used to control the volume of a stereo. The attenuator is loosely coupled to a DC motor with a speed of about 1 rpm. Loose coupling is used to prevent the motor or attenuator from being damaged should the motor be allowed to twist the control past its stopping point.

The voltage and current is controlled by a dpdt (double-pole, double-throw) switch which allows the polarity reversal when the

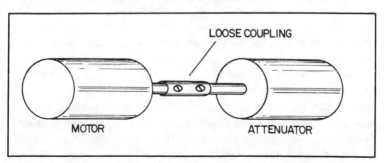

Fig. 3-1. Control of stereo volume through use of slow-turning, miniature motor at attenuator.

71

motor direction is to be reversed. To turn up the volume, the switch will be held in one position. Turning the switch in the opposite direction will reverse the motor, and the volume controlled by the attenuator will be decreased. In this manner, remote control has been gained of the volume of the stereo by using a DC motor to simulate the actions of the human hand and fingers. This rotary motion has been matched by the rotary motion of the electric motor. The reversal of the human motion is copied by reversing the polarity of the electric motor, causing the shaft to rotate in the opposite direction. By using this remote control system, the simulation of a human function has been obtained to effect the desired reaction at the work load.

DC MOTORS

A DC motor is a machine which converts direct current electricity into rotary motion. It consists of a moving coil called the *armature winding* and a stationary coil called the *field winding*. A current is passed through each of these two windings which, in turn, produces a *magnetic field*. The armature is part of the armature winding and rotates inside the field winding to produce rotary motion. The armature winding produces a rotary field while the stationary winding produces a stationary or motionless field. The motor rotates because of the interaction of these two fields. Some of the smaller DC motors replace the stationary winding with a permanent magnet which produces its own stationary field without the introduction of electrical current. The current is fed only through the armature winding when this method of motor design and construction is used. Figure 3-2 shows the interior of a simple DC motor that uses the permanent magnet to establish the stationary field. The turning axis is through the center of the rotary winding located at the two legs of the armature.

The operational characteristics of this motor are depicted in Fig. 3-3. Here, the armature is positioned so that it produces a magnetic field with the north pole at the top and the south pole at the bottom. This is shown in relationship to the polarity of the permanent magnet surrounding the armature. Unlike poles attract and like poles repel, so the north pole of the armature is drawn toward the south pole of the magnet just as the bottom south armature pole is drawn toward the north pole of the permanent magnet.

As the armature turns, the unlike poles align with north to south and south to north, referencing the armature poles to those of the magnet. At this point, all motion would stop because the poles

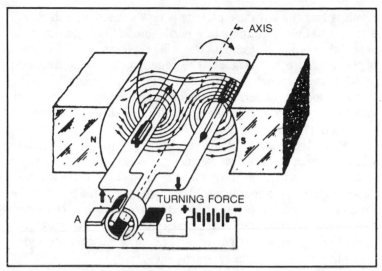

Fig. 3-2. Interior view of permanent magnet DC motor showing magnet and armature winding.

are equalized, but the motor continues to operate, because the electric current to the armature is fed through the coils by a *commutator* and *brush*. This combination allows for a complete reversal of electrical current in the armature winding by mechanically switching the DC input contacts. Figure 3-4 shows how the armature is aligned with the magnet poles when the commutation begins. The north pole of the armature is aligned with the south pole of the magnet and vice versa, but the commutator functions at this point and reverses the polarity of the armature winding. The

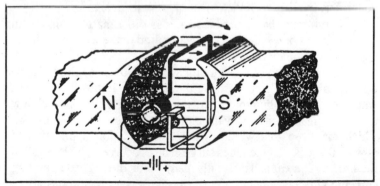

Fig. 3-3. Operational characteristic of permanent magnet motor. Armature is positioned to produce a magnetic field with the north pole at the top and south pole at bottom.

result is that the armature polarity is now aligned with its newly established field, having its new north pole in line with the north pole of the permanent magnet. The same applies to the other poles of the armature and magnet. The repelling-attracting magnetic principle is again applied until the poles are once again aligned in a north to south configuration, where the commutator once again acts to reverse the armature poles. This assures a continuous action of the rotating armature.

Most DC motors available today use several armature windings, each with its own commutator. This prevents the jerky motion of the armature which would result with a single winding, because the turning energy would be highest near the point where the poles aligned in opposite order. This turning energy or force is called *torque*, which is defined as force applied rotationally around a central axis. Torque is the motor equivalent of the force applied at the input to the lever of the rope and pulley.

The quality of a DC motor may be judged from the number of *poles* it contains. A one-pole motor has only one armature winding. A four-pole has four. Generally, the more poles and windings a DC motor has, the smoother it will run.

To understand how a motor operates, it is necessary to understand the concept of counter electromotive force (counter emf). The generator of an earlier discussion produced electricity when its armature was turned by an external, mechanical force. An electric motor does the same thing. It matters little whether the armature is turned by an external mechanical system or by the application of voltage to its internal windings. Either method produces electricity, so an electric motor is also an electric generator, but that's not the real purpose of its design.

The voltage that is produced across the armature winding in an electric motor *opposes* the voltage applied to the winding. This is counter electromotive force.

Figure 3-5 shows a DC motor with no power applied. There can be no counter emf, because the armature is not turning, but when power is applied to the armature windings, current will flow freely, being limited only by the ohmic resistance of the conductors which make up this winding. Immediately upon application of power, there is a tremendous surge of direct current. This is called the *starting current* and is usually many times the normal operating current. This vast amount of current is necessary to start armature rotation from a dead stop. The inertia of the motor armature is the cause of this phenomenon. Inertia is the tendency of motionless objects to remain motionless or the tendency of objects in motion to remain in motion. With the armature at a dead stop, it tends to stay in this configuration, and it takes more current, which can be directly related to force or power, to remove it from this natural state. Once the armature is spinning, much less current is required because of the law of inertia as applied to objects in motion. The normal state of the armature is now in a spinning mode, and it tends to remain in this state. This is the reason that total power consumption is decreased once a motor reaches maximum speed.

As the motor begins to spin, counter emf will be developed across the armature windings as they begin to revolve through the field of the permanent magnet. This, in itself, reduces the current flow, because the counter emf opposes the applied voltage to the armature windings. The faster the armature spins, the higher the counter emf becomes and the current demand is subsequently dropped even more.

The counter emf depends on several factors as to its value. First of all, of course, is the speed of the turning armature. Secondly, there's the strength of the magnetic field from the field winding or permanent magnet. A higher magnetic field strength in the field winding or magnet will result in a higher counter emf for a set rotational speed. Regardless of the strength of the field, the counter emf will never equal or exceed the applied voltage. If this were to occur, current would stop flowing, the armature would stop turning, and no counter emf would be produced.

Knowing that the higher field strengths produce greater counter emfs and that greater emfs decrease the motor's current requirements for operation, we can also realize that the greater the strength of the field winding or permanent magnet, the fewer rpm

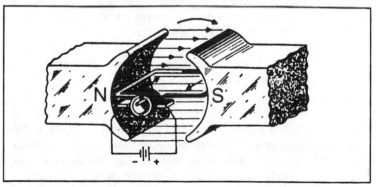

Fig. 3-4. Armature aligned with magnetic poles when commutation begins.

the motor will have to turn to produce the same counter emf. Therefore, an *increase* in field strength will precipitate a *decrease* in armature rpm. Likewise, if the field strength is *decreased*, the rpm will *increase*.

Speed is not the prime factor in motor power. Torque is. The torque or rotational force derived from a motor increases with a like increase in the strength of the field and/or the current through the armature windings. These oppose each other in one way, in that the stronger fields require less current flow through the windings to produce a certain speed. In good motor design and production, a balance is obtained which gives the best speed, torque, and power characteristics for a given motor size.

This discussion has involved the DC motor with no work load. The process described involves getting the motor up to operating speed dealing only with the inertia provided by the stationary armature. What happens when a load is attached to the armature shaft? The power demand is greatly increased. The amount of increase will depend directly upon the load. Obviously, the law of inertia will be an even bigger factor in starting current demand, because the load will probably weigh far more than the unmoving armature and, itself, will probably be in a motionless state. Now, the inertia of the resting armature and the resting load must be overcome before the motor can be brought up to speed. While a surge of current is necessary to get a motor with no load attached turning, once it reaches operating speed, it draws very little current, because it pulls no load. The only load involved would be internal friction between the armature and the bearings it rides upon. This is a very minimal value in most cases. At the same time, the counter emf produced because of the low current drain is very close to the value of the applied voltage. It may closely approach this value in motors operated in an unloaded state, but it *never* equals or exceeds the applied voltage value.

Attaching a load to the motor will cause the armature to spin more slowly. Again, the amount of reduction will depend upon the load. The counter emf is reduced due to this slowing, and the current will increase in proportion to the emf decrease. The heavier the work load is, the fewer the rpm are, and more current must be drawn from the direct current power source. If the load is continually increased, the motor will stop completely, removing *all* counter emf. This will mean that the current drain will be infinite. This is only theoretical, however, because the resistance of the armature winding will play a part in limiting the amount of current drawn in a practical motor. In actual use, though, excessive current is drawn which will quickly cause the winding to heat to high temperatures. Eventually, the windings will open up, breaking the circuit and reducing all current flow. In the meantime, however, the insulation on the windings or dust particles within the motor frame may catch fire. This is a very real hazard when large motors are used to pull heavy loads in excess of what they were designed to handle.

The DC motor discussed here is a very basic type, but it forms the principles upon which all DC motors operate. There are three basic DC motor designs, all of which may see uses in remote control applications. All of them work on the principles discussed with the main difference in each involving the way the armature and field windings are connected in relationship to each other. In the example already used, there were no field windings, their place having been taken by the permanent magnet which naturally produces an inherent magnetic field. Many of the more complex DC motors, those of high quality and considerable sophistication, will contain field windings which develop a magnetic field when current is passed through them, as was done with the armature windings in the previous example.

Series Motors

Figure 3-6 shows a schematic representation of a series DC motor. The coil representing the field winding is placed in series with the armature. Since current is the same at all points in a series circuit, the field windings and armature winding will pass equal amounts of current, both being supplied from the same source. Generally, much of what affects the field windings will also affect the armature. When the motor is in the resting state, there is no

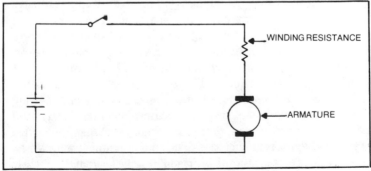

Fig. 3-5. DC motor showing ohmic resistance indication through windings.

power passed through either motor element, but when the input power is applied, high current levels are present in both windings to overcome the inertia of the stalled armature. Since current levels have a direct impact on heat buildup when encountering a resistance, the high current levels, upon initiation of power, require the field windings to be composed of very few turns in order to avoid significant resistance in the wire. The large windings contain greater lengths of copper wire and, thus, more ohmic resistance.

Referring back to the earlier discussion of DC motors, it is known that the force (or torque) increases with field strength and armature current, so the series motor offers a very high starting torque. As the motor begins to pick up speed, the counter emf, set up between the armature and the field coils, decreases the current drain from the power source, and the field strength starts to weaken. As this field becomes weaker and weaker, the motor speed increases even more, unless a significant load is present at the armature shaft to maintain a sizable current drain. For this reason, series motors should never be operated without a load or with a very light load. These conditions can cause the motor speed to quickly reach potentials which will damage the device or the very light load to which it may be attached.

From this discussion, it can be seen that series DC motors are very load conscious. The load actually determines final motor speed. With a light load, the motor will turn faster. With a heavy load, the rpm decrease. Series motors are very well suited to handle very heavy work loads for short periods of time. Because of the high starting torque, heavy doors and other opening mechanisms may be operated in a short time period by activating the motor from a dead stop. By the time the door is completely opened, the motor may be stopped without ever having reached a maximum speed. The current ratings may have been exceeded for a short time period, but in such applications, this should not create any motor damage. For this reason, the series DC motor may be preferred to other types that only develop significant torque at much higher operating speeds. The load of the door slows down the motor, because the series motor is so load conscious. This slowing automatically causes the torque to increase. In some ways, the series motor responds exactly to load condition. When the load is very heavy, the torque increases due to the motor design. When the load is lighter, less torque is provided at the shaft of the output. Using the opening of a door as a further example, once the door begins to move from a dead stop, the motionless inertia factor is overcome. The door is now moving, so this form of inertia allows it to *assist* the power source to a degree. The load has become lighter due to this motion inertia and the torque of the motor decreases accordingly, but the rpm increase. The door may open slowly at first and then increase its rate of travel as the motor increases its rpm.

The series DC motor finds many applications in remote control functions, especially where movement of load is more important than control speed. The movement of heavier work loads or the starting of a remote system which offers high inertial resistance in the first stages of activation will be especially applicable to this type of DC motor. However, systems with continually varying loads which must maintain a strict control of motor speed will find this design lacking and may be better suited to one of the two other types.

Shunt Motors

The second type of DC motor is shown schematically in Fig. 3-7 and is called a shunt motor because the field coil is wired in parallel or is shunted across the armature. A high current does not normally flow in the field windings because of this type of connection, and they usually consist of many turns of fine copper wire. Because of the size of the field windings, shunt motors do not produce a high field strength in comparison with the series motor.

Because of the field winding configuration, the current remains more constant, so a stabilized field is established upon starting which is maintained throughout the entire operation. The field will vary with load but only slightly. When power is first

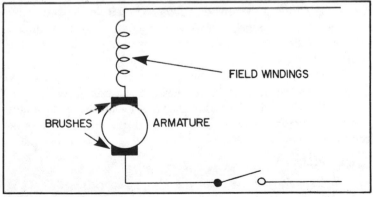

Fig. 3-6. Series DC motor.

applied to the motor, a high current will flow in the armature. In comparable motors, the current will not be quite as high as that of a series design, but the starting torque is still very good. As the rpm increase, the counter emf will approach a high level nearly but not quite equal to the applied input voltage. When this point occurs, the speed stabilizes to a near constant value.

Applying a load at this juncture of operation will begin to slow down the rpm because of the increased weight, but this creates an automatic decrease in counter emf which will cause the torque to increase. This action causes motor speed to return to a point which is near the original no-load value. Whereas, the series motor tended to make adjustments in torque for load value, the shunt motor compensates, internally, to provide a near constant speed under any load it was originally designed to handle.

Remote control mechanical devices requiring rotary motion will find the shunt motor an ideal power source if they require reasonable starting torque and may be subject to wide variations in load. Smaller constant changes in load value are quickly compensated for, and the speed remains about the same. The shunt motor is often referred to as a *constant speed motor*, because the speed remains constant over different load conditions. Load conditions may vary *greatly*, but a temporary sacrifice in speed control is the price that must be paid. This, however, is true of all motors with the shunt design providing the best speed control possible. When choosing a shunt motor as a power source for a mechanical system which tends to have widely varying loads and which may approach the maximum operating limits of the motor, it is best to choose a design which offers a higher set of ratings. This is because, as the load approaches maximum, there is a small decrease in motor speed. Operating a system that falls well within the operating curve will assure much better speed control with maximum efficiency being obtained at a point which is somewhat less than the maximum load.

To choose the proper shunt motor for a remote control application will require the exact figuring of the average work load. Once this has been done, a motor may be chosen whose maximum operating efficiency falls somewhere near the anticipated work load.

Compound Motors

We have already discussed the advantages and disadvantages of the series motor and the shunt motor. The former offers

80

excellent torque characteristics upon starting but does not offer constant speed characteristics which are found in the shunt motor. This latter design offers only reasonable starting torque. Sooner or later, a remote control system will present itself which has widely varying loads, requires high starting torque, and good speed stabilization. Neither the series nor the shunt wound motors of the earlier discussion could singly adapt to these given device characteristics. A series motor would supply the torque needed over the range of load variations, but the speed would also vary in proportion to the load. The shunt motor would provide the constant speed required, but the torque might not prove adequate. What is needed here is a combination of both the series and the shunt motor.

Fortunately, it is possible to build a motor which is a combination of the two designs already discussed. The *compound motor* shown schematically in Fig. 3-8 uses two field windings. One is in parallel with the armature while the other is in series. This combines the elements of both the series and the shunt motor field winding configuration with the single armature serving as the only element which remains constant in all of the designs.

Because of the series winding, high current will flow upon activation of the motor and excellent starting torque will be available. The speed will increase with a decrease in mechanical load but it will not increase, unlimited, because of the shunt winding setting up a parallel field which remains nearly constant throughout varying load factors. This helps to provide better speed control over what would be obtained with a series motor. It also

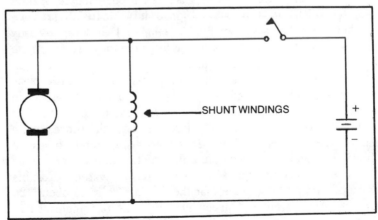

Fig. 3-7. Shunt DC motor.

prevents the motor from damaging itself internally from a *runaway* condition caused by zero or very light loading.

DC motors see very high usage in remote control systems when the desired electrical control system provides a direct current. These are relatively simple devices which convert electrical energy into rotary, mechanical energy. One of the three basic designs will fulfill most applications requirements found in remote control systems with higher rated models being chosen for the larger load factors. Another portion of this chapter will provide information on the force and power ratings of electric motors for determining the motor size in individual applications. The conversion of this rotary force will also be discussed at length.

AC MOTORS

Often, providing a separate DC power source for the control of DC motors involves transforming and rectifying the current from the AC house current line. This results in a direct current output from the rectifier circuitry at a voltage equivalent to the operating voltage of the DC motor. Of course, a certain amount of efficiency is lost in this rectifying process, because more components have been added to the power circuit. These components exhibit individual ohmic resistances to the current flow which can become significant in higher power applications.

Fortunately, this latter problem can be overcome by using motors which operate directly from the AC line or some other source of alternating current. No conversion is necessary, and the losses will be lower.

AC motors come in many sizes, each designed to operate systems which require torque and speeds it is able to deliver. They are built in a similar manner to the simpler DC motors already discussed, and the characteristics of the force they provide at the armature along with speed are also similar.

Whereas the DC motors were classified into three basic groups, AC motors fall into six categories. The main difference in these categories is in starting torque which is developed and in starting current requirements. There are many differences in the AC voltages that may be supplied to power these motor types, so the six basic designs can be multiplied by the different potentials and different phases offered by different power systems. For this discussion, we will assume that the AC line delivers about 120 volts of AC current at a frequency of 60 hertz, single phase. Commercial systems used in plants may supply three-phase voltages at different potentials which is necessary for efficiency in starting very large AC motors, but standard house current supplies the power at the 120-volt specified force in single phase and at 60 hertz, which is the power system used for discussion here. We will assume that all of the motors discussed in this chapter may operate directly from your own house current line.

As was previously stated, starting current is the determination factor (along with starting torque) for the six types of AC motors. Starting current requirements are very important, because a single-phase motor with a field coil and an armature will not start by itself. A rotating magnetic field is not developed until the armature begins to turn. Once this occurs, the field is maintained, and the armature continues to rotate until all power input is removed. Since AC motors will not start by themselves (or by the standard motor circuitry alone), some external means or an additional internal circuit must be used to begin rotation, allowing the motor circuitry to take over once the field begins to develop.

Construction

AC motors are made up of a magnetic circuit which is separated by an air gap into two, major elements. One of these elements rotates in respect to the other. The *primary windings* are usually contained by the stationary element or *stator*. The moving

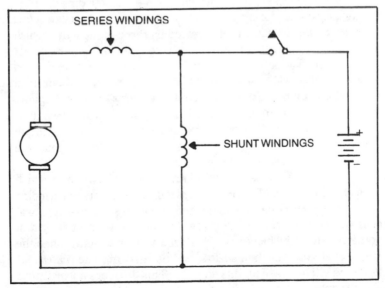

Fig. 3-8. Compound DC motor.

element or rotor contains the *secondary windings*. The former is connected directly to the AC power source. Current flowing in the primary windings will establish a field which alternates with the frequency of the line current, or 60 hertz in this case. Current flowing in the secondary also produces a magnetic field.

Single-Phase Induction Motors

Alternating current in the primary winding of an induction motor produces a pulsating or stationary field. A second stationary field is set up in the area which surrounds the secondary windings. A stationary field in both the stator and the rotor alternates with the AC line frequency, but no starting torque has yet been generated by the electrical power application. Once the rotor begins to turn at about half speed, the primary field and the induced secondary field will produce torque to bring shaft rotation up to full speed. But the problem remains of how to start the motor upon introduction of primary power. This is usually accomplished with a third winding which serves to unbalance the two fields for the period of time which is required to reach half speed. A large current consumption is usually needed to start the shaft turning from a dead stop, but once speed starts to build, the *starter winding* can be electrically removed from the motor circuit.

Figure 3-9 shows the schematic of a simple AC motor. The main winding is the only one which receives power directly from the primary line during normal operation. The rotary winding receives power from the primary winding through induction which is an interrelation between the two windings. Unlike direct current, alternating current may be transferred or transformed without a physical connection. This drawing is for theoretical discussion only because, as depicted, the motor will not begin to turn on its own due to the balance of the two fields. However, if we were to rotate the shaft of the motor by hand, sufficient torque would be produced by the winding interaction to continue the rotation and increase the rpm.

Figure 3-10 shows a practical drawing of an AC motor which uses an auxiliary winding or starter winding. If the starter winding were left in the circuit continuously, it would soon overheat and destroy itself because of the large amount of current it draws. Some way must be had to switch off the starter winding when the motor is able to maintain shaft speed on its own. The *centrifugal starting* switch is used for this purpose. This device is normally in a closed contact state, but when the motor speed begins to build, the switch contacts which are hinged to a base fly apart due to centrifugal force and break the contact. So when power is first applied to the input of this AC motor, the main field winding and the starter winding receive a current flow. The secondary or armature winding also builds up a field through AC induction. The starter winding provides the force needed to start the shaft rotating and receives current during the starting process due to the centrifugal switch being closed. Once rotation speed induced by the starter winding has reached a self-maintaining level for the motor design, this speed is sufficient to cause the centrifugal switch contacts to separate, breaking the circuit *only* to the starter winding. The starter winding is no longer used for operation.

Now suppose the primary power should fail for a few seconds, the fields of the two windings would quickly collapse and the motor would slow down. If the power comes back on before the motor has reached too low a speed, the fields will build again and the rotating shaft and armature will allow the rotation to build back to normal. But if the power should stay off for a slightly longer period, the rotational speed would fall below self-maintaining levels, and the centrifugal switch would again close due to the lessening of the centrifugal force produced by the turning shaft. When the primary power returned, the starter winding would once again be engaged to bring the shaft speed back up to self-sustaining levels before switching out for a second time when adequate speed had been built. This type of design is called a split-phase AC motor, because during the starting period, the electric current splits or flows in the starter winding leg of the circuit. For this reason, it could also be called a temporary split phase motor.

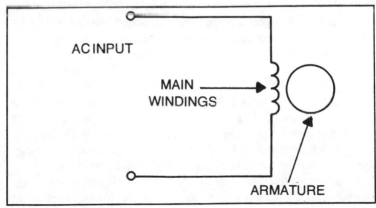

Fig. 3-9. Simple AC motor.

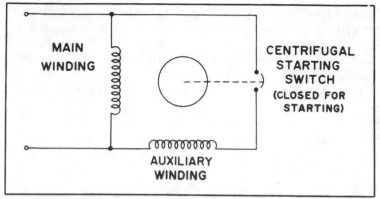

Fig. 3-10. AC motor with field winding and auxiliary, or starter, winding.

Figure 3-11 shows an AC motor design of the *permanent split phase* type. Here, the centrifugal switch has been deleted and replaced with a capacitor. A field for starting is set up in the starter winding/capacitor circuit. This provides a high amount of torque upon starting, but because of the introduction of the capacitor to the system, much lower starting current is required. Since the extra winding and capacitor remain in the circuit, the rpm are constantly monitored and increased torque is added by this latter circuit should motor speed begin to decrease. This is an excellent design, because it causes the motor to have constant-speed characteristics and requires low starting current while providing excellent starting torque.

So far, we have discussed only a few of the many types of AC motors. It is not within the scope of this text to delve fully into their operation, but it can be said that all of them work in a similar fashion using a field coil which receives current directly from the AC line which, in turn, induces another field in the rotor winding. The rotor winding is not directly connected to any other part of the electrical circuit. The force or voltage transferred through induction sets up the current flow in this latter, heavy winding which rotates with the armature. The difference in the six types of AC motors is mostly in the methods used to start them. After rotation speed has been set up which takes the motor to a little less than half normal rotary speed, the field effect within the primary and secondary windings does the rest to maintain shaft rotation. For this discussion, it is far more important to learn about motor ratings, characteristics, and rotary conversion than it is to delve deeply into the theory of motor operation.

Motor Reversing

Figure 3-12 shows a small motor with a shaft direction of left to right attached to a standard volume control or potentiometer. When power is applied to the motor windings, the shaft will slowly rotate from left to right, raising the volume level. But suppose at some later time, we wish to lower the volume using the same motor and its mechanical connections to the shaft of the volume control. To do this would require the reversal of the direction of the motor shaft turns. True, this could be accomplished with a series of gears or pulleys mechanically pushed into placed by a remote controlled solenoid or some other such device, but this complicates what should be a simple procedure. The easiest way to control volume in both directions will probably be to reverse the shaft spin direction. To do this will require knowledge of what type of motor is being used for this purpose. Some DC motors may be reversed simply by reversing the direction of current flow. The first motor discussed which used a permanent magnet instead of a field winding will do this nicely.

AC motors may be reversed by reversing the direction of the current flow in the field winding or the armature winding but not in both. In the case of the permanent magnet design, the field of the magnet is fixed and cannot be changed, but the current flowing through the armature winding, the only part of the motor which receives power from an external source, may be easily reversed by means of a dpdt (double pole-double throw) switch. This circuit is shown in Fig. 3-13. Here, the positive potential is at the upper terminal of the motor while the negative side is on the bottom terminal. The power supply, which is shown as a battery in this

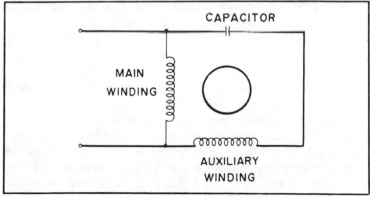

Fig. 3-11. Perman split phase AC motor.

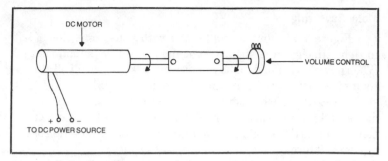

Fig. 3-12. DC motor with shaft rotation of clockwise attached to shaft of volume control.

case, is switched to the motor circuit by the dpdt device. If the motor turns from left to right using the positive on top, negative on bottom arrangement, then when the polarity is reversed by throwing the switch to the opposite position, the shaft direction will reverse. This occurs because the armature winding current which is derived from the battery source reverses, but the fixed field of the permanent magnet remains the same. Remember, to reverse the direction of a DC electric motor, the field must be reversed in one of the windings but not in both.

Figure 3-14 shows how the reversal would take place in a motor which uses a field winding instead of a permanent magnet. Here, it is necessary to isolate the power connections for the internal field coil from the DC line and connect the two leads to a dpdt switch as before. In this manner, throwing the switch in one direction will completely reverse the field produced in the field winding, thus reversing the direction of the shaft spin. It would not be possible to bring about this reversal by simply reversing the polarity of the input current without isolating the field winding and putting its input into the switch arrangement shown. As originally connected inside the motor, the reversal of input current would reverse the current flow through *both* the field winding and the armature. This would not result in a directional change.

Many of the different types of AC motors can also be reversed. Here, it is usually only necessary to reverse the field of the starter winding. Once the direction of rotation of the shaft from the starter field has been determined, the remaining windings will continue this directional rotation. Some AC (as well as DC) motors can be purchased with the direction switching contacts brought out to the exterior of the motor casing. All that needs to be done for remote control applications is to provide for the switching arrangement which is normally handled by a relay designed and rated for the anticipated current levels.

Motor Ratings

The specifications of most of the larger types of AC and DC motors are normally placarded on the motor case. These ratings will include the fram type, identification number, rpm, horsepower, volts, amperes, frequency, phase, and some other useful information. Smaller motors often are supplied only with a small specification sheet which describes the ratings in rpm and voltage. The rest is left up to guesswork on the part of the purchaser. The current consumption can be determined by inserting a DC ammeter in the line (for DC motors) during operation. AC motors may be checked as to current consumption by using an AC ammeter which may be connected externally to one of the two primary power conductors. This latter measurement is shown being effected in Fig. 3-15. The motor should be rated with no load (if applicable) to give you an idea of the general current level then with a full load to be able to rate maximum current consumption.

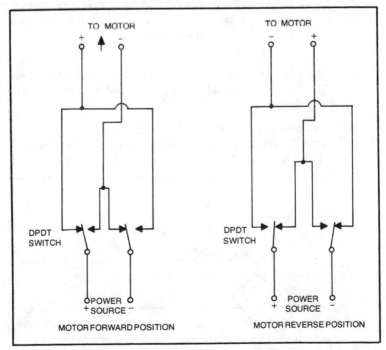

Fig. 3-13. Switching circuits for motor reversal.

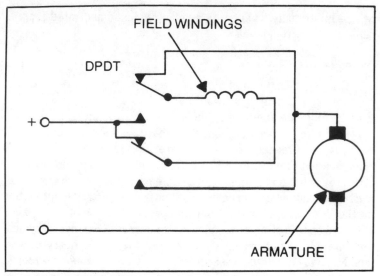

Fig. 3-14. Switching circuit for reversal of series DC motor.

Once the voltage and current are known, the total electrical power consumption may be determined by the power formula of P = IE or power equals the current times the voltage. If a DC motor draws one ampere at 12 volts, then the formula will read:

$$P = 1 \times 12 \text{ or } P = 12 \text{ watts}$$

Motors are normally rated in horsepower, especially the larger designs. Horsepower is equal to the power in watts times 0.00134, or, in formula language:

$$hp = P \times .00134$$

For this motor, the horsepower would equal 0.00134 × 12, or about 0.016 horsepower. But what does this mean? To understand

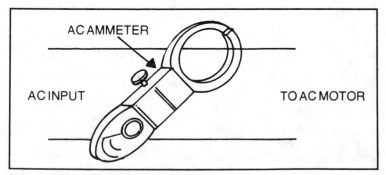

Fig. 3-15. Measuring current with AC induction ammeter.

this rating, it is necessary to convert horsepower to foot-pounds/second. This formula reads:

$$\text{Foot-pounds/second} = HP \times 550$$

The rating of a motor in foot-pounds/second equal to the horsepower rating times 550. In the example under discussion, we substitute the horsepower rating already derived into the formula for:

$$\text{Foot-pounds/second} = .016 \times 550 \text{ or foot-pounds/second} = 8.8$$

The motor will provide a power of 8.8 foot-pounds per second. Once the total force and power needed to run the mechanical system are known, it can be determined if this motor supplies the adequate mechanical power to be safely used.

Table 3-1 provides conversion of watts, horsepower, and foot-pounds/second to any of the two other units. By knowing one rating, it is very easy to use this chart in converting to another form of the same rating.

Rotary Motion

The force produced by a standard electric motor is rotary force. This is directly applicable to devices which require rotary motion for operation such as the volume control presented earlier. Often, however, this motion will have to be converted to provide a

Table 3-1. Conversion Chart for Various Power Determinations.

MULTIPLY THE NUMBER OF ——→ BY ↓ ↘ TO GET	WATTS	HORSEPOWER	FOOT-POUNDS PER SECOND
WATTS	1	746	1.356
HORSE POWER	0.00134	1	0.0018
FOOT-POUNDS PER SECOND	0.7376	550	1

means of controlling devices which cannot be directly used with rotary power.

The *capstan* was only touched on previously but is a device which can be used to convert rotary motion to *linear* motion when coupled with a length of wire or rope. Figure 3-16 shows the capstan attachment to the end of the motor shaft. Here, the rotary motion is transformed into straight-line or linear motion to possibly open a door or lift a weight when the rope is part of a pulley system. Figure 3-17 shows this latter arrangement where rotary motion is converted to linear motion by the capstan, the direction of this latter motion is transformed by the pulley, and the weight at the other end is lifted from the ground. When the capstan only winds up the connecting cable as in this drawing, it is sometimes called a spool.

Rotary motion can easily be transferred using a belt-driven system. The rotary motion of the shaft sets up the spinning of the capstan which transfers its force through a continuous belt. The opposite belt section is, in turn attached to another capstan which is spun by the transferred force. The shaft which the latter capstan is attached to begins to spin when power is applied to the motor, and this rotary motion has been transferred to a remote location, its distance determined by the length of the belt.

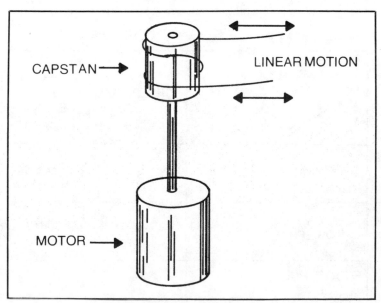

Fig. 3-16. Converting rotary motion to linear motion with a capstan.

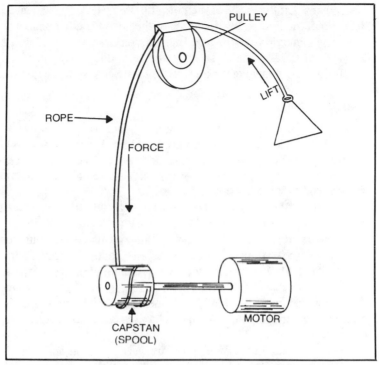

Fig. 3-17. Using a capstan and pulley for conversion of motion and direction.

In the above example, if the size of both capstans is identical, the rate of spin of the second capstan will be identical to that of the first, but the belt and capstan drive offers mechanical and speed advantages when capstan sizes are chosen for this reason.

Figure 3-18 shows a belt-drive which provides a slowing of the rotary speed by mechanical means. The first capstan is the input to the system and is one-half the outside diameter of the second capstan which is designated as the output. These capstans are usually called pulleys when used in this type of application. Because of the two to one diameter ratio, the first capstan must turn twice the rotary distance to set up one full revolution in the second. If the motor speed was 10 revolutions per minute, then the first pulley or capstan would revolve at the same rate, but the second one would revolve at half this speed, or 5 rpm. A simple formula is used to calculate speed advantage here:

$$\text{RPM-2} = \frac{\text{D-1}}{\text{D-2}} \times \text{Rpm-1}$$

This means that the speed of the output pulley (RPM-2) is equal to the ratio of the first pulley diameter to the second pulley diameter times the motor speed (RPM-1). In the given example, there was a 1:2 ratio between the first and second pulleys, or ½. The revolutions per minute of the motor were 10. So ½ × 10 = 5 rpm.

Torque Limitation

Many DC motors may have their speeds and torques controlled by a series resistor in the primary line. Figure 3-19 shows such an arrangement which may also be used with certain types of AC motors, although, in the latter, speed may not be controlled in many designs, only torque.

As the motor speed is reduced by the voltage and current limiting properties of the resistor, the torque also drops. Small adjustments to speed may be effected using this circuit, but the torque or force of the spinning shaft is also reduced. This circuit may be desirable for dropping the torque to a level where it cannot damage mechanical systems and controls should the devices become mechanically jammed for some reason. If the torque were great enough, the remote control mechanism could literally be torn apart.

The resistor in the line is of the variable type and can be adjusted for proper operation. If the correct resistance is chosen,

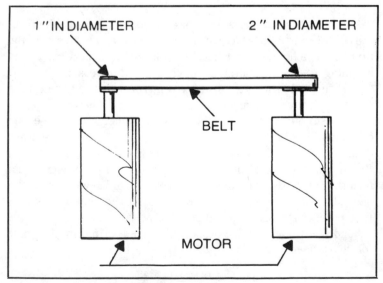

Fig. 3-18. Slowing of rotary speed using a belt drive.

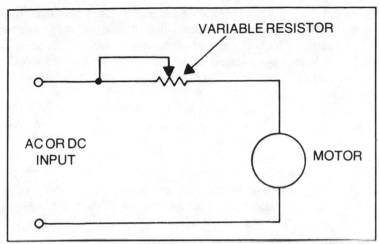

Fig. 3-19. Speed control circuit consisting of a variable resistor in series with the current.

the component will have only a small affect upon the motor and system operation. Should the mechanical device under power jam and refuse to turn, the motor will automatically draw more current, but the increased amperage will drop the voltage to a point where the motor will simply stall out. A fuse or other tripping device in the motor power line can be actuated by this current surge and the entire system shut down until the problem can be identified and corrected.

SUMMARY

Small DC motors are available from many hobby outlets and experimental catalogues and are designed for remote control applications. These devices are available in a wide range of rpm ratings and may be coupled to other mechanical devices to effect a speed decrease or increase. The formula used to determine speed reduction using the belt and capstan can also be used to increase the speed of the second pulley by making the motor driven capstan the larger of the two components, so, again, it is seen that, while the power source is instrumental to the operating of remote systems, the mechanical couplings and various machines used to transfer and to transform this power are just as important as the power source, proper.

Due to the high starting currents required by some of the large motors, it is often impossible to handle this power through the contacts of a small switch. Often, large relays and solenoids are used to switch the main power source to the motor windings. The power to these electrical remote control devices is then controlled by the small switch at the remote control position.

Power is power and force is force, regardless of in what form it is delivered. The power formulas of this and earlier chapters can often be directly applied to many different power systems. The foot-pound/second is often the most convenient level in which to reduce all power and force ratings. This is a universal term that should be used by the remote control experimenter more so than is the current practice. If this formula and all of its derivatives were used on a regular basis, perhaps, more remote control systems would work more closely to calculated specifications. It will take a little longer to run your particular system through the various formulae, but the end result will probably be efficiency in time and design.

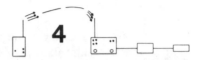

4

Hydraulic Control Systems

A brief discussion on hydraulic power systems has already been covered in this book, although the device was disguised as a water hose connected to a sprinkler head. Hydraulics are often used in far more intricate forms to accomplish work from remote points. Hydraulic systems, when properly designed, can often simulate human, motor actions and can apply a great deal of force to the work load.

Hydraulic systems are most often powered by electric motors which were discussed in the last chapter. A very real advantage of these systems is found in the fact that a relatively small motor can be used to provide a very large mechanical force at the work load. The earlier sprinkler system was a very basic form of hydraulic control. The systems discussed in this chapter are of far more practical design.

In order to understand hydraulic system operation, it is necessary to understand the basics of *pressure*. Pressure is a measurement of force exerted by a fluid per unit of area. This force is often measured in pounds per square inch and is abbreviated *psi*. A pressure measurement such as this is a quantity measurement, as pressure is not the same force. This is likened to the foot-pound per second unit discussed earlier which is a measurement of power and *not* force, which is measured simply in foot-pounds. Pressure, then, is the *power,* and the term *psi* is likened to foot-pound per second in this relationship.

The idea behind hydraulic control systems is to fill a space with a liquid. When a force is applied at the input to the chamber, a like force is present at the output. Figure 4-1 shows a simple diagram of a basic hydraulic system which is made up of a long cylinder or pipe. The input and output ends are clearly defined. A hydraulic fluid is placed within the cylinder and serves to transfer the force at the input to the output. So, if a force of 16 foot-pounds is applied at the input to this hydraulic device, approximately the same force will appear at the output. There is, of course, some loss within the system which is inevitable, but, for discussion purposes, we will assume that the input force and the output force in Fig. 4-1 are identical.

The lesson learned with the rope and pulley applies in some form to the hydraulic fluid in our basic system. In a hydraulic system with a fixed amount of hydraulic fluid, pressure will be equal at all points within the system just as the tension is always the same throughout the length of a rope under stress.

As with other forms of mechanical motion, there is a law which applies specifically to hydraulic systems. This is *Pascal's law* which states: "In a closed system filled with a fluid, the pressure will be the same in every part of that system." This is the law we continually work with when dealing with hydraulics in remote control applications. We also may use many of the other laws already discussed in making the hydraulic system perform work.

Hydraulic fluid is the carrier of the force applied at the input to the system. In the earlier example of a primitive system using a garden hose, the fluid was water. This is where the term "hydraulic" is derived. It comes from the Greek work, "hydra," meaning water. Though its roots may be in water, the modern hydraulic system used for remote control applications rarely uses this liquid. A hydraulic fluid made from special oil is normally used. Oil provides many advantages, one of which is that it lubricates the system at the same time it transfers force. Hydraulic uses involve mechanical movement, and where you have movement, you have

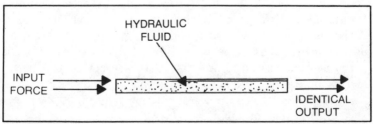

Fig. 4-1. Basic idea behind hydraulic systems is the transference of force through a contained liquid.

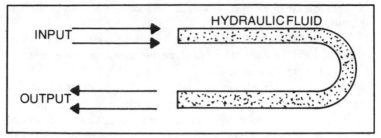

Fig. 4-2. Transforming the direction of the hydraulic force makes this simple system a machine.

friction. The oil hydraulic fluid kills two birds with one stone, so to speak.

Referring to Fig. 4-1, we see that the hydraulic system can be a machine. In this case it is not, because a machine is a device which transforms a mechanical force in magnitude, in direction, or in both ways. This example shows that the device pictured merely transfers the force at the input to the output end. No direction transformations have taken place and no mechanical advantage has been realized. Figure 4-2 shows how this system can easily be rearranged to form a true machine. Here the hydraulic hose has been bent 180 degrees. Now, a transformation has taken place. The input force which was applied in one direction has been reversed 180 degrees, so the output force is a directional transformation of the input force. Arranged in this manner, the hydraulic system of Fig. 4-1 has been changed to a true machine.

In both examples, no mechanical advantage has been gained, only force transference with the first and the addition of direction transformation with the second example. No real mechanical action of any kind has been obtained with these two examples. True, if a force of 10 foot-pounds is applied at one end, a force of 10 foot-pounds can be present at the other end, but in the examples shown, there has been nothing for the output to apply force to except the open air. Figure 4-3 shows how a piston may be added to the basic system to arrive at a machine which does provide mechanical motion. In actual use, the entire system will be completely enclosed, but, for this discussion, the output end is closed while the input remains open. Now, when a force is applied at the input, the piston will be pushed from the output with the same force as the applied pressure. A load of 10 pounds might be moved by the piston if a similar amount of force were applied at the input. This is the way all hydraulic systems operate, although most of them are far more complex than the example shown in the last drawing.

As you might have suspected, hydraulic systems can also be used to great mechanical advantage by applying some more of Pascal's law which states, to repeat, "in a closed hydraulic system, pressure is equal at all points," but, what if we increase the size of the output end of our basic system along with the size of the piston? What will be the result then? The answer is a mechanical advantage will be offered by the system which is proportional to the size of the output as compared to the size of the input. In the earlier examples, a force of 10 pounds resulted in an output of ten pounds, but if the size of the output in square inches was increased four times, the applied force to the enlarged piston at the output would also be increased four times. Figure 4-4 shows such an example.

Here, the input force is supplied by another piston which has a *contact surface area* of 1 square inch. The output piston has a contact surface area of four square inches. If the input force is 2 foot-pounds, the pressure applied to the system will be equal to the force divided by the area or 2/1. This means that a total pressure of 2 psi has been applied to the system. This is the pressure input. The formula for figuring pressure in a hydraulic system is:

$$\frac{\text{Force}}{\text{Area}} = \text{Pressure}$$

The area refers to the contact area only. This is the part of the piston which is actually applied to the hydraulic fluid or, in this case, the bottom portion only, the part which makes contact with the fluid. Referring back to the pressure measurement above, let's apply the same formula to the output which has four times the contact surface area or four inches. The formula would be a derivative of the one already discussed, but we now need to solve for force knowing the pressure to be 2 psi and the area of the output piston to be 4 sq. in.

The formula for finding force is:

Force = Pressure × Area or Force = 2 psi × 4 sq. in.

Then

Force = 8 pounds

With an input force of 2 pounds, a transformation has taken place, resulting in an applied output force of 8 pounds. This is a true machine, because the force applied has been transformed as to magnitude. As with the other systems, we did not get something for nothing. We must move the smaller piston four times as far as the larger piston is moved. Force at the output is magnified, but no

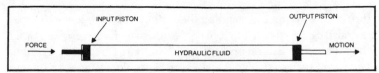

Fig. 4-3. Adding pistons to the hydraulic cylinder produces matching linear motions.

additional energy has been introduced into the system. The output measured in foot-pound's per second will be identical to the input power in foot-pound's per second assuming a perfect no-loss system.

From this discussion, it is easy to see that the first chapters on force, power, efficiency, levers, pulleys, etc. continue to apply to the more complex mechanical systems. The criteria pointed to in these chapters involve the basic laws of physics which apply to all known physical conditions. In subsequent chapters, all of these laws will be used to make practical applications out of much of the theory learned.

Figure 4-5 shows a practical hydraulic system which consists of an input piston, an output piston, and a sealed network for housing these two items and the hydraulic fluid. We know that the initial force is applied at the input and is transferred and, sometimes, transformed at the output. How is the input piston driven? Several avenues are open in this area. It can be powered by human force. By pushing the input piston with the hand, an advantage can be obtained in systems so designed. This compares with the rope and pulley system described earlier. A weight placed at the output may be easily lifted by transforming the magnitude of the applied force. Proper channeling of the hydraulic fluid reservoir can also bring about a transformation of motion. This figure shows that a downward force at the input is transformed into an upward

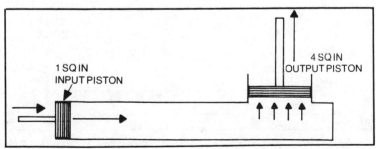

Fig. 4-4. Gaining a force advantage or multiplication through mismatching of input and output piston surface areas.

thrust at the output. Assuming the output piston to have 10 times the surface area as the input part, a force magnification of ten times the input force will be present at the output. Of course, it will be necessary to push the input piston ten times the distance the output piston travels, and the energy consumption will be identical at both points. Pushing the input piston a distance of ten inches will result in the output traveling upward a distance of only one inch. This may be a small distance, but is adequate for many remote control applications.

Figure 4-6 shows another hydraulic system. It is identical to the former example, but the input and output have been switched. Here, the advantage factor is not 10:1 as before but 1:10. This is really a disadvantage factor, because an input of 10 pounds of force will result in an output force of only 1 pound. This is a disadvantage only in regard to output force. The real advantage can be seen in the distance the new output will travel for an input push of 1 inch. A 1-inch push at the input will result in the output traveling a total distance of 10 inches. It will take 10 times the force than is seen at the output, but the total energy consumption will be the same. This type of hydraulic system is often used to control mechanical systems whose elements require a great distance of travel. This is accomplished with an input travel of 1/10th the distance for the hydraulic circuit under discussion.

This discussion of travel may lead you back to an earlier discussion of solenoids where it was stated that the latter were electromechanical devices which produced a short mechanical travel. Assuming force and power ratings were adequate, a solenoid could be used with this last hydraulic system to enable its

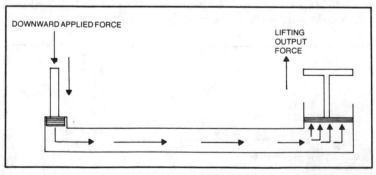

Fig. 4-5. Practical hydraulic lifting system which converts a downward force at the input to an upward force at the output. Output piston is fitted with lifting platform.

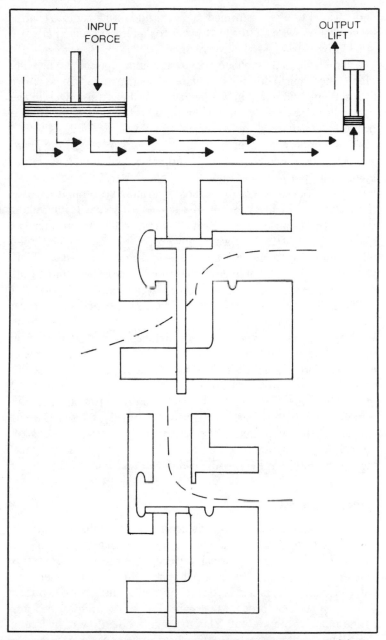

Fig. 4-6. Gaining a distance advantage by making the input piston much larger than the one at the output. A small, downward motion of the input piston produces a much longer travel in the rise of the output piston.

short travel to be transformed in magnitude into a remote control power source which ultimately produces a long travel. Figure 4-7 shows how this might be accomplished.

The solenoid is connected to the input piston. When the electromechanical device is activated, the armature will be drawn into the coil and will pull the input piston downward. Due to the hydraulic action, the output piston will travel 10 times the distance of the solenoid armature. An initial travel of 2 or 3 inches would result in an output travel of 20 to 30 inches. A powerful enough solenoid, then, could be used with this hydraulic system to effectively open and close a door by remote control.

The secret of operation of most hydraulic systems is found in the fact that they are completely enclosed. The earlier sprinkler system we discussed was a form of hydraulic action, but this was not a sealed system. The water powered the motor which turned the sprinkler head and was then removed from the system. This was a flow. A standard hydraulic system acts on flow and on the transfer of pressure within a *sealed* system. The simple systems under discussion so far in this chapter have depended upon human or solenoid mechanical power for operation. Sophisticated systems will normally use a *hydraulic pump* which is powered by an electric motor to induce the force within the fluid system. In this manner, electrical force is the main determining factor for control of the entire system and can be switched on and off in a conventional manner to allow the operator to have the ultimate control of the remote function or functions which are finally triggered by a hydraulic arm. As has been previously stated, before the control force is finally applied to the work load, it may be transferred through many different types of power systems.

HYDRAULIC PUMPS

A normal method of transferring power into a hydraulic system is through a hydraulic pump which is directly driven by a motor, usually of the electrical variety. The electrical energy is converted into mechanical energy by the motor, and this mechanical energy, in turn, is converted into hydraulic energy by the pump. The pump does not produce pressure. Rather it provides the force to the input of the system. This force sets up a flow of hydraulic fluid. The resistance to the flow is what produces the pressure.

A common hydraulic pump that is used in many different systems is the geared type shown in Fig. 4-8. This model consists

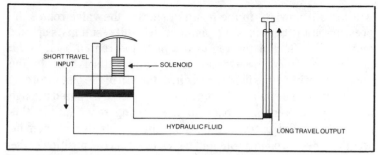

Fig. 4-7. Triggering of the input piston with a electrical solenoid. The short motion of the solenoid can produce a long motion at the rise of the output piston.

of two gears which are located in a sealed chamber. The gears are turned by the input shaft which is connected to the motor.

The spacing between the gears and the housing is very small as is the spacing between the teeth of each gear. As the gears begin to turn, hydraulic fluid is captured in the small spaces between the teeth. The gears rotate in the direction of the arrows, so fluid from the top port is transferred into the bottom part. Most of this transfer takes place between the teeth and the chamber walls, because there is so little space where the gear teeth mesh with each other. The input fluid often is obtained from a large holding reservoir whose job is to simply contain adequate fluid. When the pump is started, a positive pressure is built up at the output end while a negative pressure exists at the input.

Figure 4-9 shows a complete hydraulic system using electric motor drive. The first component in the system is the pressure relief valve which bypasses the fluid back to the holding reservoir should the overall system pressure become dangerously high and, left on its own, damage the system. This condition could result if one of the hydraulic pistons should become jammed. Next in the system comes the accumulator. The accumulator is not essential to hydraulic operation, but it is often imperative for *smooth* operation. Especially when the motor is first activated, waves or pulses of hydraulic fluid and fluid pressure may exist. The accumulator's electrical equivalent would be the capacitor which stores electrical force. The accumulator stores hydraulic pressure, so the system will start to respond immediately while the motor and pump are still building up pressure.

At this point the hydraulic delivery line directs the pressurized fluid to the directional control valve which routes to one of two hydraulics lines leading to the power piston. A fluid flow to the top will push the piston downward, while a bottom fluid delivery will force it upward. In the position shown, the valve routes the pressure at the bottom of the piston which will result in its upward motion. The fluid at the top of the piston is routed back to the holding reservoir through an appropriate return line. In the opposite position, the fluid travels in at the top of the piston forcing it downward. The fluid at the bottom of the piston is routed through a slightly different path back to the holding reservoir. In this manner the action of the piston can be changed. This may often be accomplished through a solenoid which controls the position of the directional transfer valve.

It may be a little confusing at this point, when the discussion speaks of fluid above and below the piston. This is not to mean that the induced fluid and pressure travel across the piston. If fluid is delivered at the bottom, none of it gets through to the top in a properly sealed system, but when the valve is changed, and fluid is routed through the top port, the fluid which is already in the bottom

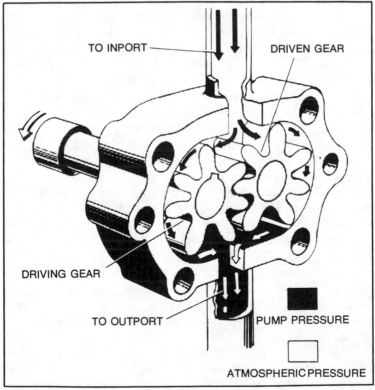

Fig. 4-8. Geared hydraulic pump.

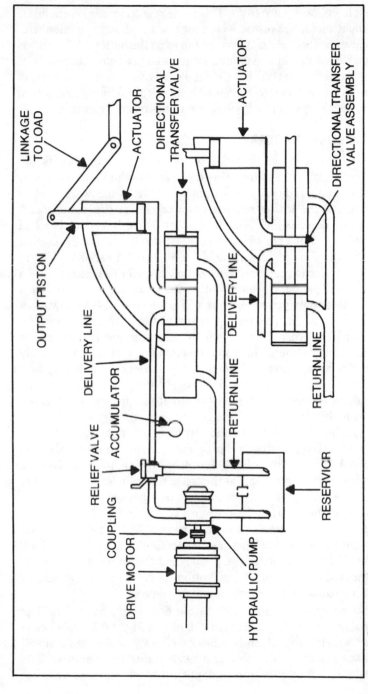

Fig. 4-9. Complete hydraulic control system.

107

must have some place to go. This is the reason for the return lines. Without them, the piston would lock in a midrange position with equal amounts of pressure on the top as on the bottom. While it has been stated that in a hydraulic system, all pressures are equal, it must be remembered, that the fluid must be continually flowing in the circuit in order to accomplish this. The delivery and return lines see that proper hydraulic action takes place at all times.

HYDRAULIC ACTUATORS

Of prime concern to the remote control enthusiast is the hydraulic arm or acutator. This is the part of the system which is applied, either directly or through levers, to the work load to accomplish the desired reaction to the original control command. A hydraulic device which provides linear mechanical motion is called an actuator. It is very similar to the piston talked about throughout this discussion of hydraulic devices. Figure 4-10 shows a simple actuator which is composed of a hollow cylinder designed to hold a tightfitting piston. This is about what a hydraulic automobile jack resembles when you take it apart. This device is also known as a hydraulic ram. In this example, the piston is forced upward by pressurized hydraulic fluid which enters at the bottom of the cylindrical chamber. This example provides working force in only one direction, upward. The piston will remain in this upward position until the pressure is released through a valve, then the weight of the piston will slowly cause it to slip downward as the pressure bleeds off.

Figure 4-11 shows another type of hydraulic actuator which allows for direct fluid control of upward as well as downward piston action. Here, the fluid inlet is chosen by a hydraulic valve. This is a two part piston with a seal at the top of the cylinder and another at the lip which is attached to the piston's bottom. When fluid is applied at the bottom, the pressure seal is at the bottom, so the piston rises, but when the pressure is applied at the top port, pressure is applied against the top portion of the piston lip (still located at the bottom of the piston) and the entire unit is forced downward. Any fluid in the bottom portion of the cylinder is returned through a line to the holding reservoir.

One type of actuator which can be directly applied to door opening and closing systems is shown in Fig. 4-12. This device uses a piston rod which is attached directly or through a simple linkage system to the door in a *single-acting* configuration. This means that fluid pressure is applied through only one opening.

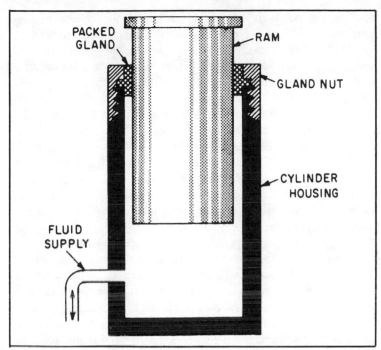

Fig. 4-10. Single-acting hydraulic ram.

When reversal is desired, an internal spring, which is isolated from the hydraulic line, returns the piston to its original position as the pressure is released. The fluid then travels out through the same port by which it entered. By observing the drawing, it can be seen that the fluid enters and is pressurized through a port located on the left side, behind the actuating piston. This forces the piston rod out, performing the opening sequence. As long as fluid pressure is maintained, the door remains open, but, when the pressure valve is opened back in the hydraulic line, the pressure on the dry side of the piston forces the fluid out of the actuator's hydraulic chamber and back into the holding reservoir. Seals are placed around the piston to prevent the pressurized hydraulic fluid from seeping past it and entering the dry area surrounding the spring. If this were to occur, pressure would be lost around the piston rod and through the air vents. These vents are necessary to avoid any resistance that might be encountered to the hydraulic push from air being compressed by the traveling piston.

Figure 4-13 shows how this actuator might be mounted to a door surface. When the pressurized fluid enters the piston chamber, the piston rod travels forward and pushes the door to its open position. When the pressure is released, the spring returns the door to its original, closed position. The spring return operation is reminiscent of the solenoid which often uses a spring to allow the armature to be returned to its unpowered, open position.

It should be stressed that the spring-return actuator actually has two, distinct chambers. One acts upon the hydraulic fluid under pressure while the other uses a mechanical spring to effect its work (closing the door). The two systems are isolated from each other by seals located around the piston head. This is called an unbalanced hydraulic actuator, because the piston rod is forced to travel in only one direction by hydraulic pressure alone. When the pressure is removed, a mechanical system in the form of the compressed spring moves the piston in the opposite direction. This latter function is not accomplished through hydraulic action.

Figure 4-14 shows a double-acting hydraulic actuator which provides true, hydraulic control of piston movement in two directions. This device has many applications in various remote control circuits. Here, two ports are provided on either side of the piston head. The shaft travels completely through the piston and protrudes from the hydraulic actuator casing at both ends. Seals are used around the shaft where it enters the piston and where it passes through the case. Seals in the piston head prevent pressurized fluid from one side of the piston chamber from passing through to the other.

Operation of this device involves the use of a hydraulic directional transfer valve, which was pictured in a previous

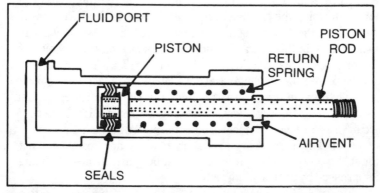

Fig. 4-12. Single-acting hydraulic actuator, unbalanced. Note internal spring for mechanical return.

110

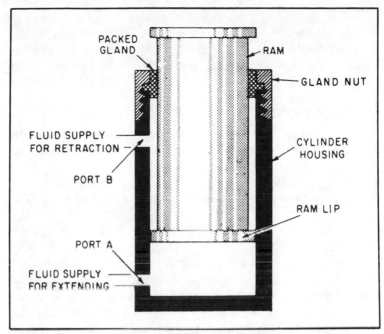

Fig. 4-11. Double-acting hydraulic ram offering two direct modes of control.

drawing. Fluid directed to the left port will result in the shaft traveling to the right. When the fluid is redirected to the right port, the shaft travels to the left, while the remaining fluid in the left chamber is channeled through a return line to the holding reservoir.

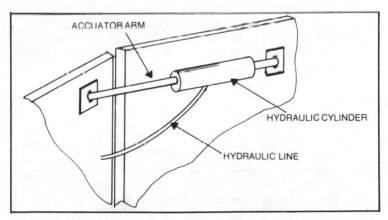

Fig. 4-13. Attachment of hydraulic actuator to door in order to effect hydraulic opening.

The double acting actuator with two ports is also made as an unbalanced device. Its diagram would reassemble that of the double-acting one, except the shaft would terminate at the piston head and pass through only one wall of the pressure compartment or chamber.

HYDRAULIC VALVES

The hydraulic valve has already been discussed and schematically shown in this chapter. These devices are many and varied, as they are designed to perform a myriad of functions. The basic valve design is shown in Fig. 4-15. This is called a poppet valve and can be activated by hand, by a hydraulic actuator, or by a solenoid. The purpose of this valve is usually to isolate one part of the hydraulic system from another. When the valve is allowed to open, the formerly unused system is actuated and performs its assigned task. The top drawing shows the valve in the closed position which has the seat fully closed. When the valve actuator is depressed, the pressurized fluid gains access through the outlet port as shown in the bottom drawing. The poppet which is the sealing ring must be constantly held in the downward position to allow hydraulic fluid to flow through the outlet. As soon as the activating pressure on the valve is released, the internal, hydraulic pressure will force the poppet back into its seated position, and the flow through the outlet will cease.

A directional control valve designed around similar lines as the above example is shown in Fig. 4-16. There is one inlet and two possible outlets. In the example at the top of this drawing, the fluid is channeled through the bottom outlet, because the poppet has sealed the top. In the bottom drawing, the poppet stem has been depressed to seal the bottom outlet port, and the flow is redirected through the top outlet.

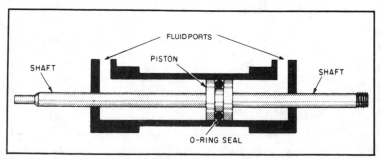

Fig. 4-14. Double-acting, balanced hydraulic actuator.

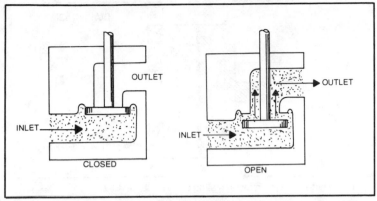

Fig. 4-15. Poppet valves showing the closed position top and the open position, bottom.

There are many other types of hydraulic valves. Some are spring loaded while others contain internal spools which are turned around their longitudinal axes to effect channeling of the fluid in two or more directions. The capacity of the hydraulic valve is totally dependent of the surface area of its inlet and outlet ports and the surface area on its internal chamber. These valves may be actuated in any of the aforementioned manners. Some are automatically controlled by the pressure levels within the hydraulic system and are designed to perform certain hydraulic switching functions when specific conditions present themselves. This is a form of hydraulic automation.

HYDRAULIC SWITCHES

In some applications, it is necessary to establish electrical control through hydraulic action. In these cases, hydraulic

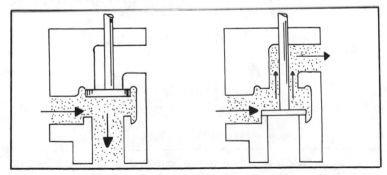

Fig. 4-16. Directional flow valve showing two paths of travel, top and bottom.

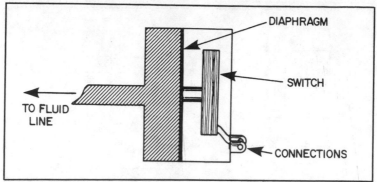

Fig. 4-17. Hydraulic switch for activation of electrical circuits.

switches are often brought into play. Pictured in Fig. 4-17, the switching contacts are isolated from the hydraulic fluid system through seals around the diaphragm. When fluid enters through the port, its pressure causes the diaphragm to travel forward, closing the contacts of the switch. An electrical circuit connected in series with the switching contacts is completed, and current begins to flow. When the pressure source is removed, a spring located within the switch, forces the diaphragm away from the closing arm, and the contracts are opened. The current flow ceases.

Figure 4-18 shows a switch which operates a bit differently on the electrical end. The former example used a separate electrical

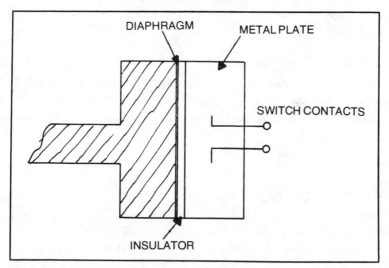

Fig. 4-18. Alternate hydraulic switch with normally open contacts in sealed chamber.

114

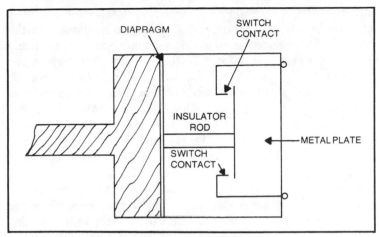

Fig. 4-19. Hydraulic switch with normally closed contacts.

switch, but this latter one makes the switch a part of the hydraulic action. Two contact points are brought out to a point near the diaphragm through insulated feedthroughs. When the diaphragm travels forward, a metal surface, which is insulated from the diaphragm, engages the contacts, and the circuit is closed. Reversing the switch is accomplished through means of a spring in the switching compartment. Figure 4-19 shows how this switch might look when set up to open contacts instead of closing them. In the normally closed configuration, the back of the switching element is closing the contacts when no pressurized fluid is in the chamber. When the fluid enters, the diaphragm and switching element travel forward, opening the switching contacts. A spring on the opposite side of the switching element returns this part to re-make contact closure when the hydraulic pressure is released.

While this discussion involves only one switch opening or closure per each hydraulic action, as with relays, many switching arrangements are available, and hydraulic operating switches may contain many contacts, some of them normally open while others are normally closed.

HYDRAULIC REMOTE CONTROL APPLICATIONS

Hydraulic systems have been used in many ways to remotely control work loads. The most obvious example is that of the hydraulic lift used in most gages in the United States. Shown in basic form in Fig. 4-20, this hydraulic device uses the ram principle or that of a single acting piston. The external head of the ram is

115

attached to the rack on which the car is placed. When the controller pulls the hydraulic valve arm, pressurized fluid is released into the bottom of the cylinder housing. A large motor and hydraulic pump maintain the pressurizing force. As the fluid flows into the pressure chamber, the ram and its load (the automobile) are forced upward. When the power altitude has been obtained, the operator returns the switch to a closed position which seals the hydraulic system. No additional pressurizing force is admitted nor is any of the stored pressure allowed to escape. The system is locked with the car in the elevated position for maintenance. At this point, no energy is being used to keep the car in this position. The pressure is trapped within a completely closed system and will be maintained indefinitely assuming there are no pressure leaks of any kind. When it is time to lower the car, the hydraulic valve is thrown into a third position which allows the pressure to slowly diminish through a check valve which also routes the hydraulic fluid back to the holding reservoir. Again, no power consumption is normally required to allow this, as the pressure which has already required the power consumption is simply being released. The weight of the automobile on the ram forces the fluid out of the bottom of the cylinder and through return lines to the reservoir. The only power consumption was required by the electric motor and, then, only to raise the car by pressurizing the hydraulic fluid for the lift. After this, the motor was no longer needed to keep the car elevated or to lower it to its original position.

Another shop and garage item is the hydraulic press shown being operated in Fig. 4-21. The device is essentially an overhead ram which is pushed downward by hydraulic fluid entering at the top end of the overhead piston. Instead of an electric motor, this model uses human power to pressurize the system. The operator drives the lever of the hand pump through muscle-power which charges the entire system. When activated, the ram travels downward, punching through the work materials placed in its path.

A highly complex hydraulic system is shown in Figure 4-21. This is a utility line truck which is used to enable electricians to work on power lines at sizable heights above the truck body. This system is far too complex to delve deeply into, but, basically, it is composed of a two part boom or a single boom which can be bent and maneuvered at the center. The base of the boom will rotate 360 degrees laterally, while the upper boom may be positioned over a wide range of altitudes from straight up on the end of the bottom boom to fully lowered.

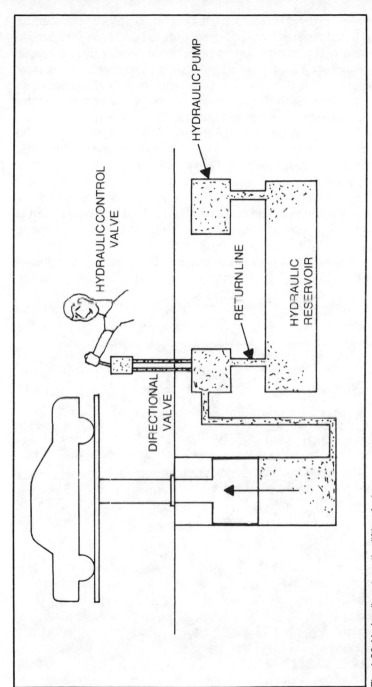

HYDRAULIC CONTROL VALVE

HYDRAULIC PUMP

DIRECTIONAL VALVE

RETURN LINE

HYDRAULIC RESERVOIR

Fig. 4-20. Hydraulic automotive lifting device.

The hydraulics are controlled either from the cab at the end of the boom or from the ground controls located near the chassis. Figure 4-23 shows the ground controls and the various hydraulic line connections, while Fig. 4-24 shows the simplistic operator controls which allow complete operation from the cab. Hydraulics are essential for this type of work, because these systems require no conductors of electricity to travel the length of the upper boom which is constructed completely from insulating fiberglass. The hydraulic lines are of a nonconducting material and the hydraulic fluid is also a good insulator. The operator in the cab, is completely isolated from any ground connections. This allows for safer operation, and should the cab or upper boom come in contact with a high voltage line, the operator is protected, because he is completely removed from any ground contacts, and current cannot flow from the line through his body to ground.

Figure 4-25 shows the operation of the boom from the ground control. The ground operator has the option of controlling the

Fig. 4-21. Operator providing power to hydraulic press by pumping pressurizer.

Fig. 4-22. Hydraulic electrical line truck showing two-part boom on bed.

entire system from this point or he can actuate a directional transfer valve which transfers control of the system to the operator in the extended cab. Figure 4-26 pictures single operator control from the boom while Figure 4-27 shows the versatility of the double boom in attaining unusual positions for almost any type of electrical work.

A relatively high-power hydraulic system powers the entire operation. It receives its charging force from the truck motor which is used in place of an electric model to drive the hydraulic pump. Should the motor fail for any reason, the system automatically seals itself allowing the operator to slowly bleed off pressure and return to the ground safely.

Due to the top-heaviness of the line truck when the boom is extended to the completely vertical position, hydraulic jacks are

119

mounted on each side of the chassis to provide the proper bracing and support to prevent swaying and possible toppling. These jacks are simply hydraulic rams with ground contact shoes which extend from each side of the chassis when activated. Figure 4-28 shows the jack being extended by the operator, and Fig. 4-29 shows the same device in the fully extended position, bracing the truck against the ground.

All of these functions involve some form of remote control, because in each case, a reaction is being obtained from a point which is remotely located from the exact work area or areas. The cab operator controls the hydraulics which are located at many points throughout the chassis and boom from a high perch, while the ground operator can control the position of the distant, elevated cab from his position on the truck chassis. The hydraulic press operator may control the device through a simple control system mounted on the hand pump which, itself, is located a fair distance from the area of the punching operation.

While all of these functions do represent remote control, they are commercial applications, and not readily usable as personal control devices. The principles by which these devices work, however, are as basic as a simple door-opener. The complexities

Fig. 4-23. Ground hydraulic controls for activation of boom and cab.

120

Fig. 4-24. Cab controls.

enter the picture when many different control functions are added to a single working machine, even though each of the many functions is very simple. It is the additive effect which give the impression of (and many times, rightly so) a highly complex systems operation.

Getting back to the more practical side of hydraulic remote control, Figure 4-30 shows a practical door opener which uses hydraulic action. Observe that the actuator is attached directly to the wall and the piston end is mounted to the door. The pump and small motor are also mounted to the wall. The actuator is a double-acting unbalanced device which requires hydraulic pressure to open and to close the door. When the motor is switched on by the electrical wiring to its circuit, it begins to turn the hydraulic pump. This pumps hydraulic fluid into the left portion of the cylinder, causing the piston to travel inward and pulling the secured door along with it. The door will remain open until the

Fig. 4-25. Operator controlling cab from ground while observing electrician working overhead.

Fig. 4-26. View of electrician in cab. Note the size of the center joint of the double boom.

Fig. 4-27. Hydraulic boom is very versatile and can be placed in a great number of positions and attitudes.

Fig. 4-28. Operating the hydraulic jack. One is mounted on either side of line truck and is forced to the ground through hydraulic action. This braces the truck and prevents toppling when boom is in extended position.

Fig. 4-29. Hydraulic jack in extended position.

motor is reversed. This pumps fluid out of the left side of the chamber and into the right side, slowly effecting the closing.

To set up this remote operation of the door, all that is needed at the remote control point is a switch which will start the motor in the direction desired. This will usually be a double-pole double-throw (dpdt) variety which is shown schematically in Fig. 4-31. The polarity of the input voltage is what determines the direction of the motor. In a practical application, the switch, which has two positions either side of the center or off position, would be labeled "open" and "close" beside each of the applicable switching positions.

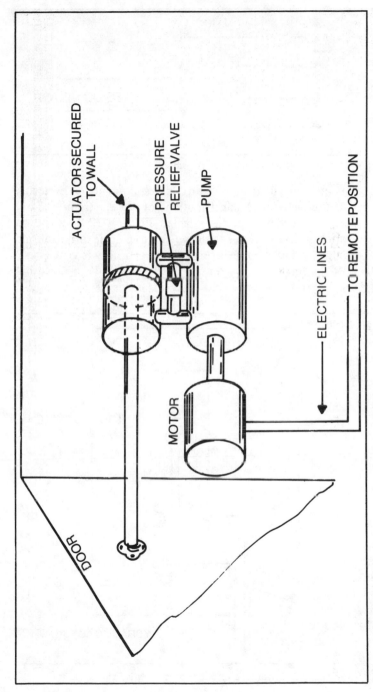

Fig. 4-30. Self-contained hydraulic actuator system for opening door.

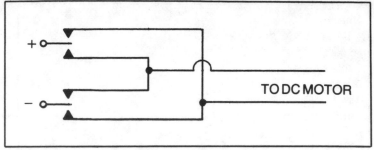

Fig. 4-31. Motor reversal switching arrangement.

An unbalanced, double acting hydraulic actuator performs many of the same mechanical motions of the electromechanical solenoid discussed in an earlier chapter. Figure 4-32 shows how a lightswitch project which used the solenoid could be turned on and off through hydraulic control. As far as practicality is concerned, the solenoid would be far better, but for the sake of discussion, this last drawing shows how it would be accomplished.

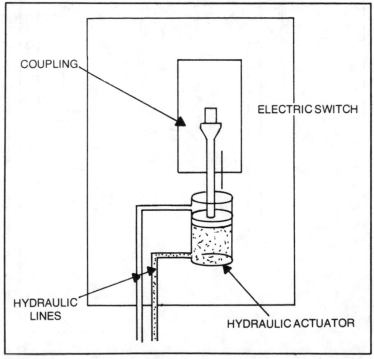

Fig. 4-32. Hypothetical hydraulic activation of common lightswitch.

The lightswitch control arm is connected by a wire or piece of twine to the actuator. The flexible hydraulic lines are run up the wall to the actuator which is mounted near the switch plate. When one compartment is filled in the hydraulic cylinder, the piston rod travels upward. This action turns the switch "on" with the arm in the up position. Reversing the hydraulic flow pulls the piston arm down to its original position, and the switch is moved to the "off" position.

Instead of running the long control lines, an arrangement might be set up which was discussed earlier in this chapter for opening a door. It used an actuator which was closely connected to its associated hydraulic pump and electric motor. The motor would be controlled by electric wiring as before. From a practical standpoint, this latter arrangement could certainly be classified as a "Rube Goldberg setup," but adequately demonstrates hydraulic control of an existing switching device.

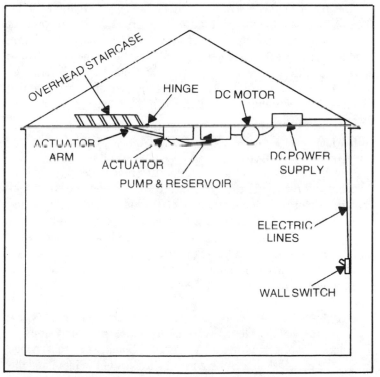

Fig. 4-33. Overhead stair lowering system showing the various components mounted to ceiling.

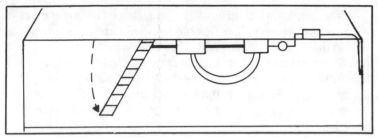

Fig. 4-34. Staircase in lowered position. Motor reversal raises it again.

Overhead staircases which must be lowered for access and raised again after use are especially adaptable to hydraulic remote control. Figure 4-33 shows how this might be accomplished. This drawing shows the attachment of the actuator arm to the ceiling and trap door on which the stairs are mounted. The motor and pump are also located nearby. When the motor is activated in one direction, the front of the hydraulic cylinder is pressurized, moving the piston backward and slowly lowering the staircase, as shown in Fig. 4-34. To raise it again, the motor rotation is reversed, and fluid flows out of the front cylinder portion and into the back. The control arm pushes out from the cylinder and slowly closes the door to the latched position. The system can then be sealed off, and the retained pressure will keep the door closed until the opening command is once more issued to the motor through the remote switching network.

SUMMARY

Hydraulic control systems offer many advantages for opening and closing hinged structures and in lifting heavy weights through force magnification. These systems are often expensive and do not readily conform to homebuilding of the actuators, pumps, and various valves. If they can be obtained inexpensively, it is relatively easy to assemble a working system from the purchased components. The hydraulic system may allow the use of a high speed motor to drive the pump without resorting to expensive gear and pulley arrangements were the motor to directly open and close a door.

Hydraulic systems operate on the principle of Pascal's law which states that in a closed system, liquid pressure is identical at all points. System pressure is measured in pounds-per square inch (psi). The physical size of the control surfaces of the input and the output will determine the mechanical or hydraulic advantage or disadvantage of the output force to that of the input.

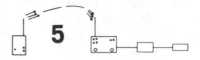

5

The Electronics
of Remote Control

Most of the discussion of remote control devices and applications has involved the wiring systems rather than the control methods. This is done purposely, because the creation and control of mechanical functions is the most difficult part of remote control. The remainder of this book deals with specific types of remote control which involves getting the commands through to the working machine. This has been accomplished in our earlier discussions by electric wiring mechanical machines such as ropes, levers, and control arms, but all of these methods required direct, physical connections of some sort between the controlling point and the working machine.

There are many other ways of getting a command signal or signals through to the mechanical portion of the remote control system without actually establishing a direct, physical contact. This is accomplished through radio signals, light beams, sound, and in other manners. This is where the electronics of remote control comes into full use. Many circuits can be made from surplus equipment already on the market from childrens toys, and from many other sources. Some may be used in present form, but most will require electronic modification to function in special applications.

In order to modify existing circuits and to build new ones, it is necessary to have a basic understanding of electronic components, circuitry, and wiring. This chapter will deal with many of the basic electronic components which will be used often and in great numbers to build some of the more sophisticated circuits offered in later chapters. Special attention should be given to the section on proper soldering techniques. In nearly 90 percent of the cases of homebuilt electronic circuits not working properly, the malfunction or malfunctions were traced to improper solder connections within the electronic circuitry. You may purchase the best components available today, but if you do not assemble them properly, they are next to useless as a circuit combination.

This discussion will limit itself to basic solid-state components which are used for remote control assembly work. If you are not at all familiar with electronics, it is suggested that you read any of the fine primers available which explain components such as resistors, capacitors, and the like. To explain all about electronic circuitry in this book would not be possible, as it would take up most of the pages reserved purely for electronic applications of remote control. This discussion assumes that the reader is at least slightly familiar with many of the standard electronic components and with the reading of simple schematics.

SILICON CONTROLLED RECTIFIERS

The silicon controlled rectifier or SCR is a solid-state device which is made up of layers of crystalline material which has been chemically treated to make it into a material called a *semiconductor*. Previously, we discussed the difference between conductors and resistors and found that one tended to conduct electrical current flow while the latter resisted this flow to a greater degree than it conducted it. The semiconductor material lies somewhere between these two as far as current conduction is concerned. In certain applications, it will freely conduct current flow while in others it will block this same flow.

Similar to an SCR is the standard silicon diode. Shown schematically in Fig. 5-1, the rectifier or diode, is made of a junction of two types of semiconductor material. This produces a device which will conduct current in only one direction. Electrical current flows from a negative source to a positive one or from a negative pole to one of positive polarity. The arrow portion of the diode schematic indicates the opposite direction to which the current is flowing with its base forming the rectifier anode and the opposite end, the cathode.

The diode allows the flow of current in only one direction and blocks it in the opposite direction. This makes it ideal for rectification purposes where electricity of varying polarity (AC) is changed into direct current. This device also has many other uses,

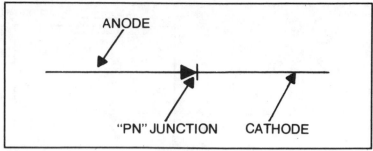

Fig. 5-1. Schematic symbol of diode showing pn junction.

but in many remote control applications, it will be found in the power supply for rectification purposes.

The real subject here is the SCR which is a special kind of diode. It allows current to pass through the device upon a command from an external signal which is applied between the anode and the third element, the *gate*. Other than this switching capability, the SCR behaves in much the same manner as a standard diode, and its output will be direct current.

Figure 5-2 shows a sample circuit which uses the SCR as a standard rectifier of AC and as a switch. The input is house current which is rectified by the switched-on SCR into the AC equivalent voltage in direct current. This is a half-wave rectifier circuit, because only one-half of the AC sine wave is acted upon by the single SCR. The switching circuit is composed of a small resistor connected between the gate and the anode. When this resistor is switched into the circuit, a small amount of current flows between the gate and anode. This causes the SCR to "fire" or, more accurately, to begin conducting current. Before this switch is thrown, the SCR appears as an open circuit and no current is delivered to the load.

Once the SCR has begun conducting current, it will continue to do so as long as the switch is left in the "on" position. When the switch is turned off, the current flow will seem to cease immediately, and for all practical purposes when using this particular circuit, it does, however, the SCR will not cease its conduction until the input current reaches zero. This occurs every 1/120th of a second in a 60-cycle AC circuit, which is used as the source of electrical power in this circuit. So, if the switch were thrown off during the peak of one cycle which the rectifier was conducting, the device would not cease conduction until the cycle completed its decay to zero.

131

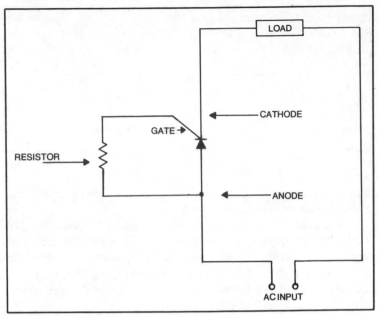

Fig. 5-2. SCR switching circuit.

This process can be more easily seen by referring to Fig. 5-3 which shows the AC sine wave as it would appear on an oscilloscope. This is a half-wave rectifier which only conducts during one half of the cycle. In this case, it is conducting during the portion of the cycle indicated by the curves on the top portion of the zero line or the positive cycle. During the other half of the cycle, it is blocking the flow of current. Now, if the switch were to be thrown during the part of the cycle marked "X," the SCR would not cease its conduction until the cycle peaked and then returned to a zero value. This lag time of less than 1/120th of a second usually means nothing in remote control applications as long as alternating current is used as the power source. This demonstrates, however, that once conducting, the SCR will continue to conduct until its current flow reaches a near zero value.

Perhaps, it seems that we have dwelled too long on this tiny fraction of a second of cut-off time for the SCR, but this becomes all important when we switch to a DC power source and use the SCR for control and switching under these circumstances. Figure 5-4 shows the same basic circuit, but this time, instead of an alternating current input, direct current is substituted. Since a rectifier will pass the flow of current in only one direction, it will easily conduct DC which is of a fixed polarity. This assumes that the diode is connected with correct polarity observation as is done in Fig. 5-4. To reverse the diode would result in a zero current output.

As before, a small resistor and switch set up a circuit between the gate and the anode. When the switch is thrown, the SCR will fire and current will be conducted through the load. Now, it comes time to turn the current off. Throwing the switch to the off position will have no affect whatsoever. Why? Because the SCR is conducting and will continue to do so until the current it is passing drops to near zero. The AC circuit did this several times in a fraction of a second, but DC stays at its constant value at all times of operation. Result: The SCR will continue to conduct, regardless of the position of the gate-anode switch, until the source of current is removed from the circuit. Then, the SCR will quickly return to its, static, nonconducting state, and it will be necessary to turn the gate-anode switch to the "on" position to return it to a state of conduction once again.

TRIACS

It can be seen that the SCR is a true rectifier in all aspects and its output will always be direct current. This is alright if the load needs or will operate from DC, but what if it is desirable to supply alternating current to a load? This can be accomplished by combining two SCRs in reverse parallel. Figure 5-5 shows how this is done. Two, matching SCRs are placed in parallel as shown. The anode of one is connected to the cathode of the other and the cathode of the first is connected to the anode of the second. The

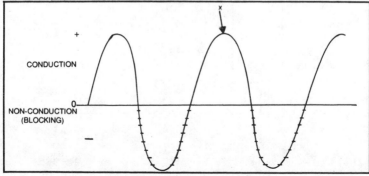

Fig. 5-3. AC sine wave showing the conduction cycle and blocking cycle in half-wave rectifier circuit.

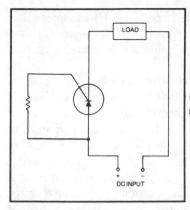

Fig. 5-4. SCR circuit using DC input supply.

gates are tied together. Now, current of one polarity will be conducted through one SCR, and when the current reverses, it will be conducted through the other. Both diodes are located within the same major circuit, so alternating current will be passed by this combination device when both are triggered. Fortunately, it is not necessary to buy two SCRs and build this nonpolarized circuit yourself. It is available as a single, compact component about the size of a single SCR. The circuitry is connected internally through a chemical semiconductor manufacturing process. When made in

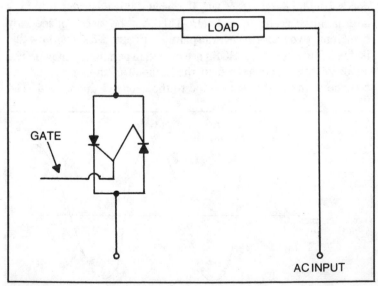

Fig. 5-5. Two SCRs may be combined as shown to pass current in either direction.

134

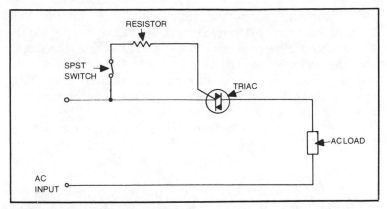

Fig. 5-6. Simple triac circuit for AC loads.

this fashion, the device is no longer called an SCR but takes on the name of *TRIAC*. A triac is a semiconductor device which, when triggered, will pass current in either direction. It is ideal for control of AC circuits.

Figure 5-6 shows a simple circuit that uses the triac to control current flow through an alternating current load. Again, a small wattage high value resistor is used as the triggering mechanism in combination with a small single-pole, single-throw switch. The triac appears as an open circuit when in its nonconducting state, but when the triggering switch is thrown, conduction is begun. As with the SCR, the triac will continue to conduct until the current source reaches zero. Fortunately, in AC circuits, this occurs every 1/120th of a second, so the switch may be turned off and the circuit will cease conduction within an instant. The switch must be left closed at all times of operation for the same reason the circuit is so easily shut off. If the switch is thrown off, the conduction quits in a scant fraction of a second.

Triacs are normally quite inexpensive and can be purchased to handle large amounts of current. They are often used in AC motor speed control circuits. Triacs and SCRs alike have certain current and voltage limitations. The former are self-sufficient when it comes to survival, because if the voltage level is too high, they will usually simply cease to conduct, whereas an SCR may be destroyed.

Figure 5-7 shows how an AC source might be controlled to an AC load using the triac and a remotely operated relay. The relay simply replaces the spst switch. The spst contacts are contained on the relay. When the relay is activated by closing a remotely located switch, current flows in the main circuit because the triac will conduct. The relay could be of a small DC-actuated variety to avoid the running of higher voltage control lines. Of course, the supply could be controlled directly by the relay in series with the main line, but the triac has no moving parts and is more reliable. Then, too, it can be placed in a much smaller area than can even a small relay. The control or switching contacts could be brought out of the circuit to the relay which might be mounted in an external position from the semiconductor control circuitry. While a relay was used in this example, the actuator could easily be a hydraulic switch or another type of remote control system.

The last drawing used only a standard relay which means that the circuit must be activated at all times the triac is to be conducting. An easy power-saving modification would be to add a latching relay in place of the one shown. Now, it will only be necessary to supply a command pulse of current to allow the relay to move to its latched position. The circuit is now self-sustaining until another remote pulse of current opens the relay circuit and ceases triac conduction. This latter design is most often used in practical applications.

TRANSISTORS

A transistor is a device which is most often used in electronic circuitry today. They can perform a myriad of electronic functions and are the most versatile solid-state devices known today. Figure 5-8 shows how a relay may be controlled with a transistor switching circuit. When a small signal is applied between the emitter and the base, the transistor begins to conduct current through the collector and the emitter. Unlike the SCR which is also a DC device, when the signal is removed from the transistor input,

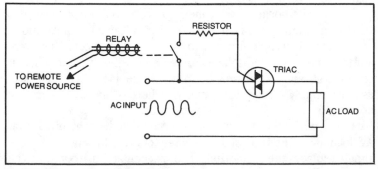

Fig. 5-7. Relay control of triac circuit.

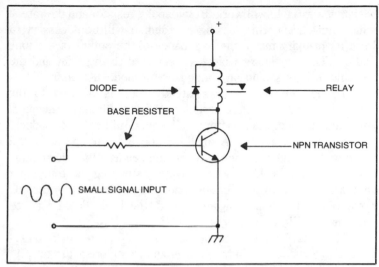

Fig 5-8 Transistor control of DC relay acting on small input signal to base.

the transistor ceases conduction and returns to its normal, off state.

A transistor is similar to two diodes connected at the cathodes. The cathode junction is the base, while the anode terminals serve as the emitter and the collector. This applies only to the bipolar transistor which was the type used in the previous circuit. There are two types of bipolar transistors in common use: the npn and the pnp. Both are shown schematically in Fig. 5-9. Transistors are used for switching purposes as in the previous circuit example, for amplification of power, and for many other electronic applications. The designations of bipolar transistors have to do with the types of semiconductor materials used to make them, and more accurately, in the ways this material is layered or "sandwiched" to form the finished transistor.

While all of the solid-state components discussed so far in this chapter required an input current, the transistor was the only one which specified an input signal instead of deriving its triggering current from the main current source. Either the SCR or the triac may receive their triggering pulses in the same manner, although proper precautions must be taken to make sure this separate triggering circuit is isolated from the current flow path of the major supply source. Figure 5-10 shows an SCR circuit which can be triggered from an audio tone. Here, the fixed resistor has been replaced by one whose resistance can be varied. Almost any high resistance variable will work. Between the resistor and the gate, a small audio transformer has been added. It will be necessary to have its primary match the impedance of the output of the tone source. The secondary which is connected to the SCR and the variable resistor should have impedance of about 1000 ohms.

To operate the circuit, a steady tone is applied to the transformer input with the resistor set for maximum resistance. After this is done, the resistor is adjusted until the SCR conducts and the load is activated. (A small light designed to operate at the value of the applied voltage is recommended as a load during this test). Once the load has been activated, stopping the tone input should cause the SCR to cease conducting at the end of the AC cycle. When the tone is again applied, the load should activate again as the SCR, once again, conducts.

This latter circuit can be used for direct remote control applications using a radio transmitter as the remote actuator. The tone generating device can be a receiver with a bfo (beat-frequency oscillator) tuned to the transmitter frequency. When the transmitter is keyed, the received carrier will be changed into a tone at the audio output which is coupled to the SCR conduction cycle and the load receives current until the transmitter is turned off. Here is true *remote control* without the necessity for elaborate hookup wiring or hydraulic lines as would be the case with some of the methods described in earlier chapters, and this system could have the remote transmitter located miles from the actual work load.

An alternate system could use a receiver without a bfo, and the tone could be generated at the transmitter input. Either way, a tone is relayed from the receiver, through the matching transformer, and, finally, to the SCR where conduction occurs. The reaction is obtained in the load performing its desired function.

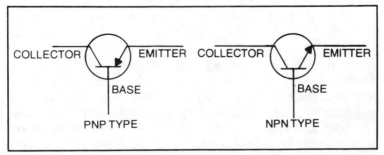

Fig. 5-9. Schematic representations of bipolar transistor; pnp type and npn type.

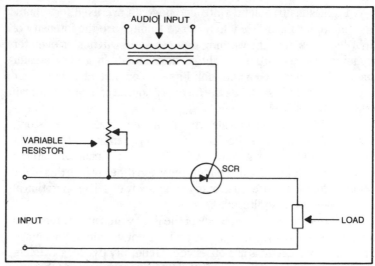

Fig. 5-10. Audio input drive of an SCR switching circuit.

This same system applies to the triac which can be connected to an AC load and to the tone source through the matching transforming in the same manner. Varying audio outputs from the tone source can be used by readjusting the variable resistor. Depending on the load current demand this entire circuit could be constructed at home for less than ten dollars even if you had to buy every part at a retail price. Fortunately, everything but the triac or SCR can be salvaged from an old tube-type radio receiver, although the miniature tranformer may have to be stolen from a transistor pocket portable. It can be seen that this remote control circuit offers many advantages and is far simpler than the mechanical systems described in earlier chapters. Again, the mechanical systems are usually much more difficult as to design than are the electronic circuits to be discussed throughout the next chapters of the text.

LIGHT-SENSITIVE SOLID-STATE DEVICES

Solid-state devices have been on the market for many years now and we are all familiar with the transistor, the diode, and even the integrated circuit. Some devices which are not as familiar includes the light-emitting diode (LED), the phototransistor, the solar cell, and the cadmium sulfide photocell. A light-emitting diode gives off a glow when it passes current while the other devices react in some way, shape, or form when outside light strikes their sensitized areas.

Light-sensitive solid-state components are used for a number of applications. They may serve to measure the intensity of light level, as daylight warning devices and switches or even for communications purposes. These devices act in a very similar manner to more common solid-state components, but their triggering mechanism takes the form of a source of light rather than tone or current flow.

Semiconductor materials which make up diodes, transistors, and other solid state devices are very sensitive to nuclear radiation. All forms of radiation will affect semiconductor material in some way or other. The conductivity and other properties of the semiconductor are altered in many complex ways. The operation of these devices is also altered.

Depending on the frequency of the light which may extend into the infrared and ultraviolet bands and are not visible to the human eye, different reactions are observed. Generally two major effects occur when light strikes a semiconductor material; one of these is called *photo-emissive*. The second effect is called *photo-conductive*. A photo-emissive device would include the solar cell which actually generates electrical current when light strikes its sensitive surface. A photo-conductive component would include the cadmium sulfide cell whose resistance to electron flow is changed by the amount of light on its surface. A photo-emissive can be thought of as active while the photo-conductive device is passive.

By using these devices, circuits can be made which provide power from converted sunlight. Other circuits will contain their own power supplies while an external source of light controls the various electronic functions. Many interesting circuits can be made using one or many light-sensitive solid state devices. Hobby circuits such as code practice oscillators, radio receivers, and radio transmitters take on a whole different aspect when light sensitive devices are integrated into their circuitry.

SOLAR CELLS

Photoelectric cells are photo-emissive devices which respond directly to the presence of light. Often called solar cells, they can be used to directly power a circuit requiring low voltage and current. They usually consist of a flat plate with a specially treated surface, although many other shapes are also found. Solar cells can be treated almost like dry cell batteries, as they have a positive and a negative terminal and can be connected in a parallel circuit for increased current or in a series circuit for more voltage output.

Figure 5-11 shows a solar cell device. Normally their output is around 0.45 volt. Until recently, solar cells large enough to charge batteries and to operate electronic equipment of moderate current demands were very expensive. Technology, however, has brought the price of these units down so that even those with moderate power outputs are well within the range of the experimenter.

The output from a solar cell is dependent upon the amount of light which is present at its sensitive surface. This light energy is transformed into electrical current. Greater light levels or intensities will create higher current ratings. After a certain point, an increase in light intensity will have no noticeable effect because the cell is operating at its maximum designed output. Most hobby-type solar cells will provide adequate current when subjected to artificial light from an incandescent bulb or even a common flashlight. When many solar cells are combined in series and parallel combinations, enough current may be generated to charge small 1.5 volt batteries or power remote control systems.

On the commercial market, solar cells are used to power AM transistor radios, digital calculators, watches, and some very expensive units are even used to supply emergency electrical power in place of an automobile battery. The solar cell has been available to experimenters for a much longer period of time than have the other light-sensitive components to be discussed.

CADMIUM SULFIDE PHOTOCELLS

All materials used to fabricate electronic components offer resistance and conductance to the flow of electrical current.

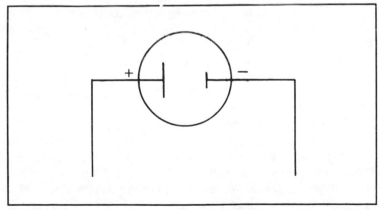

Fig. 5-11. Schematic representation of photovoltaic or solar cell with polarity indicated.

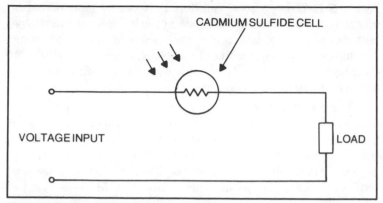

Fig. 5-12. Cadmium sulfide cell in simple circuit.

Devices which hinder the flow are called resistors while those that pass the flow are known as conductors. A special semiconductor device that exhibits a resistance to the flow of electrical current is called a cadmium sulfide cell. It can be thought of as a variable resistor much like the volume control on a radio or television; but instead of a shaft which is turned by hand to control the resistance, the cadmium sulfide cell depends upon light rays striking its sensitized surface to control the current flow. When no light is present at its surface, the cell exhibits its highest resistance to current flow. As light intensity is increased, the resistance becomes lower and lower until it finally reaches a point of *minimum resistance*.

The cadmium sulfide cell is thought of as a passive device which is photo-conductive in nature. Its conductive effects are directly related to the amount of light which strikes its surface. This device is normally used in circuits which have another source of electrical power. Unlike the photocell, the cadmium sulfide cell produces no electrical current on its own but controls the flow of electrical current already present in the circuit. Sometimes a cadmium sulfide cell will be used in connection with a photoelectric or solar cell to form a circuit which is not only powered by light intensity but controlled by it as well.

Cadmium sulfide cells are used for control through placement in the base leg circuit of a bipolar transistor to turn on and turn off state of the latter device. When light strikes the surface of the cell, the resistance changes, varying the amount of current delivered to the base lead. The transistor will then conduct or change to another state. This circuit can be used to trigger a relay or activate other

142

types of electronic circuits. Figure 5-12 shows a circuit which contains its own power supply. When light strikes a cadmium sulfide cell, the transistor conducts and completes the circuit through the relay which is used to activate other pieces of equipment.

LIGHT-EMITTING DIODES

A light-emitting diode contains semiconductor material which has been treated with the chemical, gallium arsenide. When current is passed through these devices, they emit light in the infrared as well as the visible ranges depending upon the type of materials used and the manner in which they have been treated. LEDs come in many different sizes, shapes, and forms and are used mainly as indicators for electronic devices. LEDs take the place of the old panel lamps which drew much more current.

The glow which emanates from an LED is a cool light and is not produced through heating effects. Owing to this, LEDs are much more efficient as to energy consumption than are other types of artificial light. LEDs are often available in integrated circuit forms which will display numbers and letters when current is fed to the proper contacts.

While LEDs are not, technically, light sensitive solid state devices, they are mentioned here because they are often used with the true light sensitive components already discussed to form triggering circuits. In this application, one or more LEDs are placed so their light rays will strike the sensitized surface of a solar cell or cadmium sulfide cell. Here a conversion takes place several times: DC current is passed through the LED and converted to light which strikes a photocell and is converted back to DC current or the light from the LED strikes the surface of a cadmium sulfide cell which, in turn, controls the flow of current in a separate circuit. Figure 5-13 shows a typical light-emitting diode with its two leads that must be connected with respect to polarity.

PHOTOTRANSISTORS

Photocells are composed of a junction made when two types of semiconductor materials are pressed together in such a manner that when light strikes near a junction, an electron flow is released. If this transparent junction is backed up by another semiconductor

Fig. 5-13. A light-emitting diode.

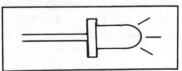

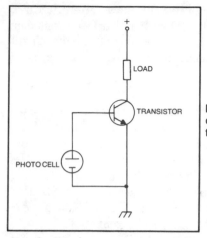

Fig. 5-14. Photocell control of relay circuit from the base of the transistor.

layer called a collector, photoconductive current generated by the photocell junction is amplified as in a transistor. The sensitized area becomes the base junction of the transistor and when light strikes this area, it produces an input signal to the device. This is very similar to a circuit which can be formed using a photocell and a pnp transistor. This circuit is shown in Fig. 5-14. The phototransistor does away with the separate photocell or photodiode portion of the circuit and houses it in one unit.

SPECIAL SOLID-STATE DEVICES

The three basic devices detailed in this chapter are the SCR, the triac, and the transistor. Only brief theoretical information has been provided, because it is more important for our purposes to know what they can do and what they can be used for rather than to fully understand their principles of operation. These three devices have many derivatives, one of which is the light-controlled version of each.

Light activation is important to remote control applications where a beam of light may be used as the main controlling force. Modern science has removed many of the problems from these applications by developing SCRs, triacs, and transistors which can be directly triggered by a ray of light. Other than their triggering methods, these electronic components operate in the same manner as the more conventional types already discussed.

Figure 5-15 shows a representative circuit of a LASCR (light-activated silicon-controlled rectifier). Notice that it has the three main elements as before, but the schematic shows two arrows pointing to the schematic representation. This indicates light activation. The device will not conduct when the window or light collector mounted on the device is in the dark, but when light strikes it, the SCR will trigger. Removing the light will cause it to stop conducting when the input current reaches a near zero value as before. The light-activated triac operates in the same manner.

Figure 5-16 shows a light-activated transistor or phototransistor. This looks just like the schematic of the bipolar transistor except the base lead is often removed. The base or biasing input to the phototransistor is in the light rays which are played upon the light collector. Light intensity can be made to cause the component to conduct electricity. In other applications, the transistor can form the input to an audio oscillator which will vary in tone as the light level increases or decreases. With all three light-activated devices, light rays can be used as the prime control element.

Some of these light-sensitive solid-state devices may respond only to light rays of specific frequencies (infra-red, ultra-violet, etc.) and can be chosen from the dealer or manufacturer to perform according to the specifications of the system they are to operate in.

Light control is discussed in another chapter, but it is important to remember that, other than by their unusual triggering methods, light-activated solid-state devices cannot be differentiated in operation from their more standardized counterparts. They even appear physically similar to these other devices.

Another solid-state component which has become readily available at reasonable prices is the light-dependent-resistor

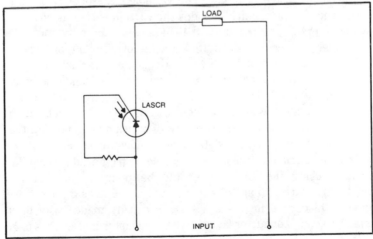

Fig. 5-15. LASCR light-activation circuit.

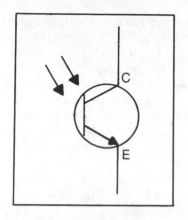

Fig. 5-16. Light-activated transistor (phototransistor) shown schematically.

(LDR). This is the popularly used name for the cadmium-sulfide cell. This is made from semiconductor material which tends to conduct electricity when exposed to light. The LDR has two leads and does not require electrical connections which observe polarity. It may be used in either AC or DC circuits interchangeably. Unlike the other devices discussed, it does not require a power supply from any external source. Pictured in Fig. 5-17, the LDR changes its internal resistance which is dependent upon the light which strikes its collector on the top of the case. In the dark, the resistance may be almost a million ohms (or more, depending on the device design), but in bright light, it may show a resistance of less than 10 ohms.

This latter device is shown in a remote control circuit which also includes a standard SCR. The variable resistor between the anode and the gate (which includes the LDR in series) is adjusted so that the SCR does not fire with the ambient light in the room. At this stage (depending upon the amount of ambient or normal lighting), the LDR is in a mildly resistive state. When a light ray is beamed upon the optical window, the internal resistance instantly goes to a low level. The current flow within the gate-anode circuit increases due to the lower resistance and the SCR triggers. If an AC supply voltage is used, when the light beam is turned off, the SCR will quickly revert to its static, nonconducting state.

While a triac or transistor may also be triggered in a similar manner using the LRD, it should be pointed out that the light-activated versions of each of the three, basic solid-state devices discussed in this chapter are usually made up of their standard, semiconductor layers to which an internal LDR has been added.

146

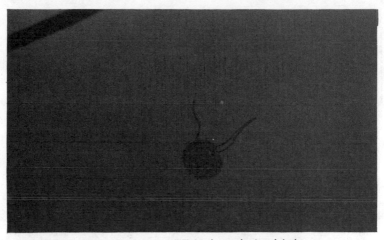

Fig. 5-17. Cadmium sulfide photocell (light-dependent resistor).

Of prime concern to many experimenters and remote control enthusiasts is the photovoltaic or, as it is more commonly known, the solar cell. Pictured in Fig. 5-18, this solid-state component is the only one which actually produces power. More accurately, it transfers the power from another energy source, light rays, into an electrical source or system. When light strikes the surface, some of the energy contained in the rays is converted to electrical current. The output from most solar cells is about 0.45 volt or a little less than a half volt. They may come in a wide range of current levels which are dependent upon the device size. The larger cells usually offer a higher current output.

Solar cells can be used like batteries. They can be wired in series circuits to increase the available voltage. Such a circuit is shown in Fig. 5-19. Thirteen cells have been connected in series to deliver an output of about 6 volts. Due to the series connection, the current rating of a single cell must still be observed at the higher voltage. So, if one cell would deliver 0.45 volts at 400 milliam-

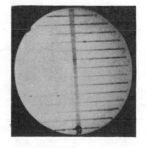

Fig. 5-18. Photovoltaic or solar cell.

147

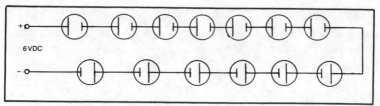

Fig. 5-19. Solar cell series circuit.

peres, a bank of 13 cells will deliver about 6 volts, again at 400 milliamperes. Figure 5-19 shows a parallel connection of the 13 cells. Here, the current is multiplied. The voltage at the output will still be 0.45 volts, but the current rating will be 5200 milliamperes or 5.2 amperes. In either circuit the same amount of maximum power is delivered to the load in bright sunlight. Figure 5-21 shows a bank of solar cells used to form a triggering source in a sun-activated automation system. This supply delivers an output of 6.3 volts at about 3 watts.

Since solar cells generate their own electrical power, they can be used to directly power tiny motors designed to run within the limits of the solar cell's output power. These motors could be used to turn miniature controls and receive their power through the controlling source. Here the activation energy (light rays) is directly converted to the main power source which is then changed to mechanical or rotary power through the motor.

For automatic control systems, solar cells or any of the other light-activated devices can be set up to switch on a remote system when the sun rises. Many dawn to dusk lights operate with circuits which use these components except the relays they control are normally closed when the light-activated components are in the dark. When the sun rises, the relay is released.

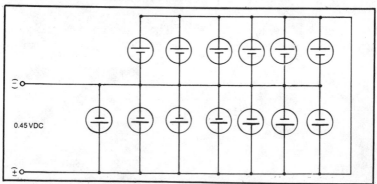

Fig. 5-20. Parallel circuit comprised of solar cells.

148

Figure 5-21 shows a circuit which uses a photovoltaic or solar cell in order to trigger a relay when light strikes the pick up surface of the cell. This device then supplies a small amount of current to the base of a transistor which conducts and causes a relay to activate. If you should desire to have the light beam turn off a device instead of turning it on as is the case with this example, simply substitute the relay with a similar one containing normally closed contacts. Using this latter circuit the load or device under power will remain on as long as the photocell surface is in the dark or in dim light, but when light strikes its surface, the transistor will

Fig. 5-21. A bank of 13 solar cells.

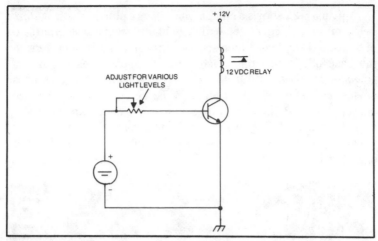

Fig. 5-22. Adjustable control of relay through variable resistor placed in solar cell/base circuit.

conduct and the relay will activate, breaking its normally-closed contacts.

Figure 5-23 shows a similar circuit which is activated in a slightly different manner by using a light-dependent resistor. The absence of light causes the relay to activate. A light beam directed on the LDR surface plate will cause the relay to disengage. Alternate contact closures can be used as with the previous example to reverse the switching affect.

Either of the two basic circuits discussed here could be used for remote control of a work load by using a light beam from a strong flashlight. The chapter on light-activation explains these circuits in more detail and provides suggestions on focusing the light beam for more reliable activation techniques. The main problem with light-activated controls is the affect ambient or natural light has on the control mechanism. Should natural lighting intensity increase, the circuit may have to be adjusted to take this condition into account.

ACCESSORIES

When using light-sensitive devices, it is often important to shield their sensitive areas from certain types of light which emanates from other than one, specific source. Shields have been designed which fit around the components and channel light rays from a specific direction onto the surfaces of the various devices. Others may block all light, acting as a shield until removed for circuit operation. Still others act as filters which block certain types of light from the components while allowing others to pass.

Remember, when using many of these devices, any type of light will act as a driving signal. Just as it is necessary to shield certain mechanical switches to avoid their accidental activation, it is also necessary to shield light activated-switches to prevent the same occurrence. You must think in terms of "light-activation". When building these circuits, you must realize that, in most cases, light is the main element which begins, ends, or establishes circuit operation.

TRANSDUCERS

A transducer is a device which takes one form of energy and converts it to another form. A microphone is a transducer as it turns the energy which is contained in audible sounds into a current flow in an electrical circuit. A stereo speaker is also a transducer. It takes current flow in a circuit and converts it back into audible sound. Photo-emissive, sometimes known as photovoltaic devices, are also transducers. They convert the energy from light into electrical energy which assumes the form of current flow in an electrical circuit. Photo-conductive devices, such as the cadmium sulfide cell or light dependent resistor, are not tranducers in the direct sense, because they do not convert energy found in light sources but simply react to it by changing their internal resistance.

This explanation of transducers will better help you to understand and to realize that light is a genuine and very tangible

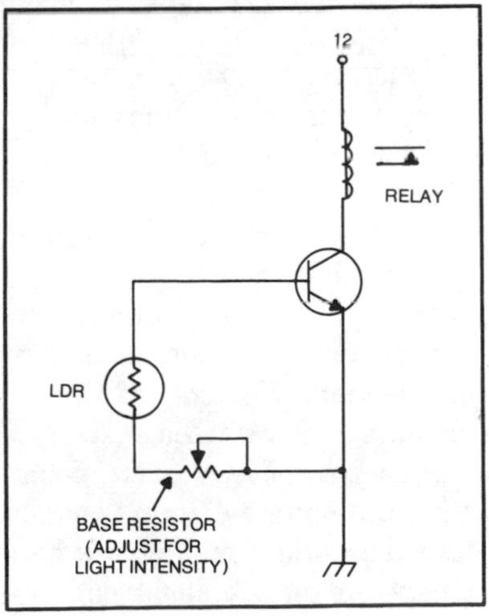

Fig. 5-23. Relay control by placing LDR or cadmium sulfide cell in base circuit of transistor.

source of energy even though it is usually thought of as an intangible. The light-sensitive, transducer devices simply transfer this energy form into another form which can be more easily measured and utilized.

INTEGRATED CIRCUITS

Years ago, transistors made the scene in the electronic world and were raved about as being the answer to a whole new phase of electronic technology. This statement was absolutely correct but at the time few people realized that the transistor and other solid state components would lead to the development of an even more progressive and useful electronic device which would rev-olutionize the electronics industry and take monumental steps in the direction of electronic miniaturization. This electronic device is known as the integrated circuit (IC).

The integrated circuit is normally composed of the types of materials which are used to manufacture discrete devices such as transistors, diodes, thyristors, etc. In fact, all of these devices may be found in an integrated circuit. The term integrated circuit aptly describes the device. An integrated circuit is simply a complete circuit which is integrated onto a single chip of semiconductor material, usually silicon. Whereas a transistor is a single, discrete device when purchased as such on the electronic market, an integrated circuit will have many transistors or other solid-state devices along with resistors and other components, all housed in one compact unit. Indeed, most integrated circuits weigh a fraction of an ounce, can be held between the thumb and forefinger and may actually contain over 1000 solid-state devices.

The main requirement of an integrated circuit is that all of the components such as transistors, resistors, and diodes be proces-sed from the same solid state material. Several techniques are available for the manufacture of integrated circuits, but their construction is basically the same as the forming of other solid-state devices which use a silicon material. This is treated with impurities to form a semiconductor. The various types of semiconductor materials are combined to form the different components. The chip of silicon used in an integrated circuit is simultaneously treated or stamped, so to speak, in a manner which will cause various microscopic portions of the water to be transformed into different components. A small part may act as a transistor while another will be a diode or resistor. Many of the circuit contacts brought out of the case to a connection point.

Owing to this, an integrated circuit may be wired to form several different circuits depending on the wiring configurations used. Figure 5-24 shows an example of an integrated circuit as it would look if discrete components were used to form the same circuit. A circuit made with the discrete components would be considerably larger than the size of the integrated circuit, which is about the same size as one discrete transistor. The enormous saving in physical space is responsible for the compact size of many types of very sophisticated remote control equipment manufactured today. The integrated circuit can be treated almost as a discrete component when combined with other portions of a larger circuit. For example, one integrated circuit can be used as a limiting amplifier in an automatic gain control circuit which is incorporated with other circuits to make up a shortwave receiver. Instead of thinking in terms of transistors or diodes, we are now thinking in terms of completed circuits.

Another great advantage of using integrated circuits whenever possible is dependability. In the days when the vacuum tube was used almost exclusively in electronic circuits, dependability was of prime consideration in the size and almost armored construction of all types of equipment. This was to protect the fragile glass tubes from shattering should the equipment be accidentally bumped or dropped. When transistors and other such devices came along, a great deal of this concern for circuit damage was avoided due to their basic construction. However, wiring connections between discrete devices would break upon severe impact, and while the circuit dependability was at a much higher

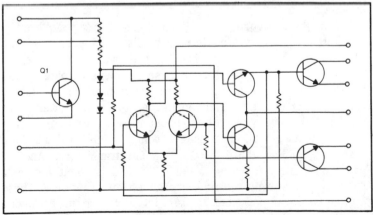

Fig. 5-24. Schematic equivalent of single integrated circuit chip.

level, failures often occurred when the electronic equipment they used was subjected to a high degree of stress or vibration.

By using integrated circuits, whole circuits have become as mechanically dependable as a single transistor of a few years ago. If several integrated circuits can be used to complete circuit without using external discrete devices, the dependability is even higher, and maximum reliability is achieved when a single integrated circuit is used with a minimum of external connections or devices. ICs are impact resistant for most practical purposes, not usually subject to vibration damage, and can be incorporated into a small, compact enclosure which offers the highest degree of component protection.

Another great advantage of the integrated circuit over circuits which are built using discrete components is the cost factor. An integrated circuit which can be bought today for less than $2 will completely take the place of the same circuit using discrete devices costing over $500 at today's prices. The inexpensive nature of the integrated circuit makes it highly desirable for use in any modern electronic construction. Many electronic circuits can be completed using integrated circuits at a cost of 1/100th or less of what the cost would be using discrete devices.

It can be seen that integrated circuits offer many, many advantages by using modern construction techniques. While discrete devices are still used in many types of remote-control equipment, the integrated circuit is fast replacing them just as the transistor and diode replaced the vacuum tube years ago. Modern technology calls for the use of integrated circuits whenever possible to continue to develop the state of the art of electronic building.

It might be thought that with all of the advantages offered by integrated circuits, there might be some disadvantages. This is not really true, although integrated circuits are usually more subject to damage from heat, created when these devices are soldered into the electronic circuit. Additional care is warranted when installing ICs but other than this, their application to the circuit are very similar to that of any discrete component. Integrated circuits are available in many different packaging configurations which include cases resembling those used for small transistors. Another packaging arrangement is called a *dip* and is identified by a rectangular case with contacts emerging from either side. These are the two most prevalent forms of integrated circuit packaging and are shown in Fig. 5-25. Special sockets are available for the dip packaging, and allow all soldering to be done with the IC out of the circuit to prevent any damage from the heating effect of the soldering pencil.

To summarize, the availability of high-quality integrated circuits at low prices is very important to remote control applications. For about the same price of a transistor a few years ago, integrated circuits can now be used which contain many transistors, resistors, and diodes.

ELECTRONIC WIRING

When putting together any electronic project, certain skills and tools are necessary to complete the job in a manner which will assume correct and lasting operation. Many projects may be designed for installation on perforated circuit board material which is available at low cost from almost any hobby store. This type of construction allows for detailed assembly and provides a sturdy and compact base for all circuit components. Figure 5-26 shows a piece of perforated circuit board. The wire leads from the components are simply pushed through a convenient hole in the circuit board. Wiring connections are made on the bottom of the board by twisting various wire leads together or connecting them by using single pieces of hookup wire and then soldering. This will also provide for a very neat appearance on the top of the board where the components are located and a sturdiness which is not easily obtained with point to point wiring on contact strips.

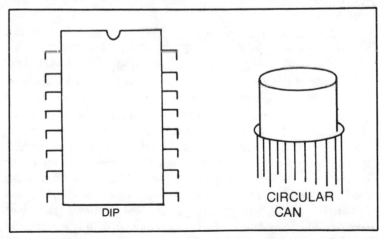

Fig. 5-25. Integrated circuit packaging. Dual-inline package is on left. Circular can is on right.

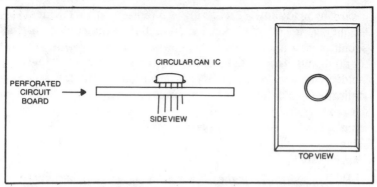

Fig. 5-26. Mounting an integrated circuit on a piece of perforated circuit board.

Several ways of mounting components can be used and the procedure followed will depend largely upon the amount of space available on the circuit board and also on the type of circuit being built. Figure 5-27 shows examples of vertical and horizontal mounting of various electronic components. Vertical mounting takes up less horizontal space and therefore requires a smaller section of circuit board. Horizontal mounting the components will require more horizontal space but will not require as much vertical room and provides a flat finished circuit. This latter type of wiring is ideal for circuits which must be housed in a box or container. Vertical mounting will generally take up the least amount of space because the circuit will not be spread out as far.

TOOLS

When building the projects outlined in this book, a normal assortment of shop tools will usually be all that is needed. Included in your complement should be needlenose pliers for bending the wire leads of the components and for wrapping these leads with others at the same contact points. Diagonal wire cutters will also be needed to cut these leads to the proper lengths and for trimming after soldering has been completed. Additionally, a pocket knife is desirable which may be used to cut sections of perforated circuit board and to scrape insulation from the surface of painted or enamel-coated wire. A pocket knife can also be used to clear away blobs of solder which may accidentally drop on circuit board material and other components. It is also good to have a wire stripper, electrical tape, and an assortment of Phillips and flathead screwdrivers of the small to miniature variety. Epoxy cement will be found useful for securing the bulkier components to the circuit board to prevent movement from vibration. Other tools which might also come in handy but which are not absolutely necessary would include a set of nutdrivers for installation of the circuit board in the metal case and a desoldering tool for removal of improper solder connections. Alligator chips will be useful for protecting delicate components during the soldering operation. These should be purchased in a general assortment of many different sizes. A type of insulation known as heat shrinkable tubing can also be used. This is fitted around the bare leads of components which are in danger of being shorted out to ground. This loose fitting insulation is tightened to the conductor by applying heat from a soldering iron. Upon heating, the tubing shrinks and molds itself closely to the conductor.

SOLDERING PROCEDURES

Soldering is a most important part of assembly for any electronic project regardless of the type of circuit or the components used. One of the largest manufacturers of electronic kits has stated that almost 90 percent of the failures involved in putting these kits together has been traced to faulty or improper solder

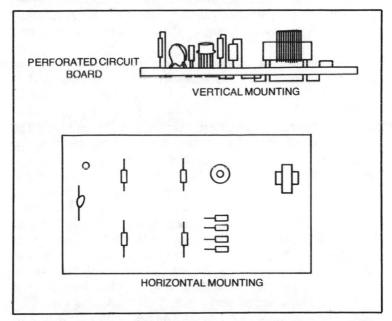

Fig. 5-27. Vertical component mounting on circuit board and horizontal mounting.

contacts. A hasty job of soldering electronic circuits will result in a circuit which does not perform properly or dependably or one that may not function at all. Soldering is always a critical part of electronic circuit construction and must be done with care while using the proper technique.

The soldering iron most desirable for assembling the projects in this book is the pencil type which has a power rating of about 30 watts. This provides enough heat to get the job done, but it does not get so hot that the fragile components are damaged. Soldering guns are very popular for certain types of electronic assembly but most of these are rated at 75 watts or higher. Most of these guns provide temperatures which are much higher than required for the assembly of the smaller projects. Also, the soldering tips of these guns are overly large for some of the compact applications. Some manufacturers offer expensive soldering stations which include a pencil soldering iron, an insulated holder, and a control box which keeps the temperature of the iron constant at all times. While these devices are very convenient for these applications, they are not necessary, and a simple soldering pencil from a local hobby store may be purchased for less than $10. Make certain you follow the manufacturer's instructions when preparing the tip of a soldering pencil for its first use. This is usually when the tinning procedure takes place and involves heating the iron and applying a small amount of solder to the tip.

Only one type of solder is suitable for use in the construction of electronic circuits. This is known as resin core solder which is sold by most electronic hobby stores and is always identified as such. There is another type of solder which may be sold in hardware stores and plumbing outlets which has an acid core. The center of this solder contains an acid which is desirable for plumbing applications but which will ruin electronic circuits and components. The corrosive acid core solder usually results in cold solder joints in electronic circuits, and the acid will gradually eat away at the delicate electronic components. A cold solder joint is a connection which has not been made properly and results in a high resistance. High resistance joints present most of the problems in improperly soldered electronic circuits. These joints do not adequately conduct the flow of electrical current and can some-times cause rectification in audio circuits. A cold solder joint is a poor or absent electrical connection. It is most often caused by simply dropping the solder onto the joint before the elements have been heated to the proper temperature. This can occur when the tip of the soldering iron is applied to the *solder* rather than to the *joint* to be soldered.

Proper Soldering Techniques

There are several steps involved in forming solder connections when building electronic circuits. Each of these steps must be followed in order and to the letter in order to arrive at a completed product which is electrically stable and dependable. The steps are as follows:

☐ Make certain the elements to be soldered are clear of all foreign matter or debris. Wire conductors for example, should be scraped clean of all insulation and wiped free of oil, tar, or grease.

☐ A firm mechanical joint must be formed from the elements of the joint before soldering is attempted. This is done by tightly wrapping the conductors in such a manner that no physical movement is possible between elements.

☐ The soldering iron should be of an adequate temperature to allow for proper heating. It should be turned on a few minutes before soldering is attempted.

☐ The soldering iron is applied to the joint, *not* to the solder. Once the joint has been properly formed mechanically, the soldering iron tip is placed against it to allow it to heat to the same temperature.

☐ The solder is then placed against the joint, not against the soldering iron, and allowed to flow freely around the elements. When the joint is heated to about the same temperature as the soldering iron, its elements will meet the solder and allow it to flow into every part of the wrapped conductors and contacts.

☐ Apply only enough solder to get the job done. Too much solder can create a cold solder joint.

☐ Once solder is flowing in the joint, remove the tip of the iron and make certain that the joint is not allowed to move. Movement at this point may cause the cooling solder to become cracked or loose in certain areas of the joint.

☐ Allow about 20 seconds for the solder to cool.

☐ Wiggle the protruding elements of the joint to make certain that no physical movement occurs where the solder bond has taken place.

☐ Examine the appearance of the solder joint, looking for any signs of a dull surface or globular solder deposits. A proper solder joint will have a smooth, shiny appearance while a dull, rough appearance indicates a cold solder joint.

While these steps may sound complicated upon first reading them, they will become second nature to you as you complete more and more solder connections. After only a few hours of practice, it will take only seconds to solder each joint in an electronic project. The main trick to soldering is to always apply the tip of the iron to the joint and not to the solder and to always allow the solder to flow into the crevices of the joint before removing the heat. Remember to use the least amount of solder necessary to get the job done. Cold solder joints result when too much solder is used because the cooling rate is different for different layers of the molten lead which is applied to the joint elements. A soldering iron applied to the outside of a large blob of solder may cause the outer portion to become molten while the inside remains relatively hard. This inside portion is normally the part of the solder joint which performs the electrical bonding.

A firm mechanical joint is stressed, because solder is not of adequate mechanical strength to form this physical type of bonding. It serves only as an electrical bond, not as a mechanical one. If solder is used to hold two connections in place, for example, normal stresses will cause this connection to work loose and the solder contact to crack if proper mechanical rigidity of the joint was not originally established. Again, solder forms an *electrical* joint. The elements of the joint must form the *mechanical* connection.

Even when using the low wattage soldering pencils, speed in making the joint is often very important. Many of the solid-state devices used in projects in this book can be damaged or destroyed when they become heated past their maximum points of endurance. If you are not experienced in proper soldering methods, you would do well to practice upon a more rugged device such as a resistor, capacitor, or even upon two wire leads wrapped together. Practice proper soldering techniques until they become second nature to you. This will increase the speed with which you're able to make the joint. This is important because the longer the soldering iron is applied to the leads of a component, the hotter the component gets. A happy medium must be reached wherein adequate time is taken to complete a good solder joint without taking so much time that the components become excessively heated.

A *heat sink* is used to aid in the further protection of heat sensitive electronic components when soldering. This is a device which *sinks* or absorbs heat. A pair of needlenose pliers can serve as a very good sink when used to tightly squeeze a lead at a point near the shell of the component. Heat will travel up the lead from the point where it is being soldered, but the larger mass of the needlenose pliers will absorb most of it which prevents a great deal of heat from reaching the case or shell of the component. Alligator clips and special heat sink clips can also be used to form a good source of heat protection. These devices have the advantage of remaining in place after the clip contact has been made and will free the builder's hands for other parts of the soldering procedure. When applying a heat sink, make certain that it is not located too closely to the point on the lead which is being soldered. Placed too closely, the heat sink can pull heat away from the joint and create a cold solder connection. The heat sink is best placed at a point on the lead nearest the component case.

Again, the proper soldering of electronic circuits is of paramount importance in electronic building. If you take shortcuts when putting together the circuits in this book, you are bound to run into trouble either when the circuit is first tested or, later, when poor soldering connections break down. A few minutes spent in properly completing a project can save future hours of troubleshooting, resoldering, and replacing heat-damaged components. Do not attempt to even start on a project until you know the correct methods of soldering.

GOOD BUILDING PROCEDURES

In addition to the proper soldering techniques just discussed, there are certain procedures that should be followed when building any type of electronic circuit. These are designed to help the new and inexperienced builder become more proficient at what he is doing, to cause himself the fewest problems, and to successfully complete all of the projects he attempts to build. We have all seen results of half completed projects. These are the ones which were started long ago and which were to be completed as soon as an additional part was obtained, but which just never got finished. Half-completed projects are often subject to breakage and other types of damage because they are usually not installed in a box or covering which adds protection.

The uncompleted project is not usually the result of lack of interest, lack of ability, or lack of skill. It is often the result of beginning a project before the builder had all of the parts necessary to complete it. This is a cardinal rule of electronic building. Never begin a project until all parts, components, connectors, and the housing are on hand to complete the project. When you begin an electronic project with certain components missing, you cannot build the circuit in an orderly manner as would be the case if all components were on hand. The builder makes certain mental notes about parts which have been left out and which are to be replaced and then, at a later time, completely forgets about them. A few of the components which were not on hand originally may be obtained, wired into the circuit, and then, assuming that the project is finished, power is applied. Unremembered by the neophyte builder, a component or two was omitted from the circuit, a component which the builder made a mental note of. Since he has forgotten all about this, he assumes the circuit is finished and finds that it does not work properly or at all. He now has to troubleshoot the circuit and will be more apt to look for poor connections or damaged components rather than to seek out *missing* components. Often the circuit winds up a total failure and is tossed in the junk box as a source for spare parts. This is a good example of a circuit which probably would have worked perfectly if proper techniques had been followed.

While many builders will complete an electronic circuit having all the electrical components on hand, the case or box which is to house this circuit is often saved for last. There is nothing quite so fragile as an electronic circuit on a perforated circuit board which is not mounted in a protective case. These circuits have nasty habits of falling from work benches or of breaking when accidentally placed under heavy objects. As soon as your circuit is completed, it should be immediately mounted in a protective case after initial testing.

A major cause of improperly wired circuits is fatigue. Experienced builders never work on circuits when they are tired, sleepy, or when their minds are on other things. If you work too long on a small circuit board, your vision will often start to blur and your hands may begin to shake from being in one position for too long. When you feel the least bit fatigued, stop what you're doing and take a 10- or 15-minute break until you feel refreshed again. Don't set a specific day or time to have your circuit completed. If you begin to run behind, you may start to rush or work past your point of adequate concentration. The only result from this will be a circuit which may have possible problems due to polarity reversal of components, wiring errors, or improperly formed solder joints.

Make certain that the work bench area where you assemble your circuit has adequate lighting and ventilation for ease of construction. Arrange your seating so the normal assembly of circuits will not put you in an uncomfortable position, causing you to strain or reach in such a manner that you tire rapidly.

Most of these suggestions for good building techniques are merely common sense ideas and should be obvious to most individuals. It is a good idea to have one specific area where electronic assembly is normally done. This gets the builder accustomed to working under set conditions and mades him more comfortable and relaxed.

By following these assembly suggestions, you should be satisfied each time a project is completed, both with the quality of your electronic circuit and with its operation and dependability. You should also take pride in the fact that a great many electronic projects which are built by other individuals not adhering to these techniques are going unfinished or are causing problems when completed.

SPECIAL INTEGRATED CIRCUIT BUILDING TECHNIQUES

All of the heating effects which create problems in building with integrated circuits can be overcome by using a socket. The socket is soldered into the overall circuit before the device is inserted and there is no possibility of any damage occurring by heat. Adequate time may be taken when soldering these sockets without fear of any type of heat damage. Proper soldering techniques are still dictated, however, as a cold solder joint at a socket will cause just as severe a problem within the circuit and possibly more because of the added resistance created by the friction contact of the device within the socket.

If solid-state devices are to be used with sockets, it is extremely important to make certain that the device leads are cleaned of any foreign materials, especially those of an oily nature. A dirty lead can form a high-resistance contact within the socket and cause the same types of circuit problems which are most often related to cold solder joints. The socket contacts too, should be cleaned to make certain that no grit or foreign material has covered the areas which make contact with the leads. Periodic inspection of the sockets is necessary, especially if the electronic circuit is used out-of-doors or in an area which is subject to dust and dirt. A circuit which uses sockets is not quite as dependable as one which uses direct solder contact, so if mobile or high-vibration applications are anticipated, the socket technique may not be practical.

COMPONENT MOUNTING

Integrated circuits used in this text are of two varieties, the circular can and the dip which is an abbreviation for Dual In-Line Package. There is a third integrated circuit configuration which is called a flat pack. This last type is most often used for computer applications and is very difficult to work in a typical home shop due to the extremely close spacing of the circuit leads.

Circular can integrated circuits are the type which look very much like transistors with many leads instead of just three. Often, a small tab will protrude horizontally from the bottom edge of the case to give some means of reference when determining the pin connections of the IC leads. The mounting this type of integrated circuit to a circuit board is identical to the mounting of transistors except more device leads must be contended with. This packaging is most conducive to the home builder because it allows for point-to-point wiring and does not necessarily regulate the builder to using printed circuit boards. When using the DIP integrated circuit, circuit boards are always required unless a socket is used which terminates in long wire leads instead of the normal pin connections. Wire lead extensions can be soldered directly to a dip IC but the chances are great that this process will damage the component because of heating effects. It is almost impossible to connect any sort of heat sink to the extremely short pins. Also, this packaging is usually accomplished with a plastic case and will melt and become disfigured under conditions of extreme heat. The use of a dip socket will alleviate all of these problems.

Packaging actually has played a great role in the development of the integrated circuit. The IC was first developed to be used in electronic flight control systems for space vehicles. While transistors were useful for this purpose, their physical size prohibited them from being used extensively. Initial space flights could be rated at costing well over one million dollars per pound of material lifted into space. Obviously, purely from a cost standpoint, the miniaturization of even small transistor circuits was essential. This desire for a smaller package was the impetus for the development of the integrated circuit. Another contributing factor to the development of the integrated circuit has been the overwhelming use of computers, not only by the United States government but by businesses, large and small.

The mounting of integrated circuits and other solid-state components when building a remote control project in this book is best accomplished by using a small piece of perforated circuit board which is available at most radio and hobby supply stores. It is also recommended that sockets be used for any integrated circuits which are available only in dip configurations. Figure 5-28 shows how an integrated circuit of the circular can variety can be easily

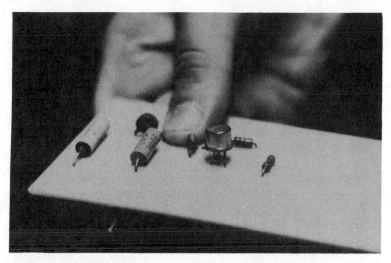

Fig. 5-28. Placing integrated circuit leads through vector board.

mounted on this type of circuit board by inserting each of the leads through a separate hole and then soldering from beneath. This also creates an attractive, neat looking finished circuit while being viewed from the top side of the board. Many of the circular can packages for ICs contain a small plastic tube at their centers which acts as a divider that keeps the package slightly above the circuit board. This allows for adequate ventilation on all sides of the device housing. This type of construction is, technically, point-to-point wiring as opposed to printed circuit board construction, but the perforated circuit board acts as an excellent base or mounting platform for all components.

Figure 5-29 shows how a completed circuit might look. Notice that a vertical mounting technique has been used to conserve space for the resistors involved in the structure. This is accomplished by bending the top lead of the resistor down along the side of the carbon body and clipping both leads so that the ends are even. The same is true of the mounting of small electrolytic *capacitors* which contain axial leads. These same components could just as easily have been mounted in a horizontal position (flush with the circuit board) if so desired. The vertical construction is intended to conserve horizontal or circuit board space. All connections are made from beneath the board by twisting various leads together and soldering them in the correct fashion.

Integrated circuits of the dip variety will often fit in a perforated circuit board with closely spaced holes. Point-to-point wiring may be used with this type of IC if it will fit the circuit board properly but an IC socket would be preferred. It is important to use care when installing a dip IC in a socket. The pins of the integrated circuit are very delicate and are easily bent or even broken when forced improperly. Correct insertion procedures dictate that you line all the pins up on one side of the IC with the holes along one side of the socket. Notice that each pin is tapered in a manner which suddenly becomes square at the midway point. Now, start each pin into its own socket hole but do not seat them all the way. In other words, only the tip of each pin is started into its respective hole in the IC socket. Next line up the pins on the other side of the IC into their respective socket holes. Make certain that the pins on the first side have not slipped from their holes while this is being done, it may be necessary to slightly bend some of the pins in order to get them to align properly. This can be accomplished with a toothpick or other small pointed device to gently force the tip of the pin into the correct slot. At this point, check all of the IC pins to make certain they are correctly inserted into each of the socket slots. Now, press firmly at the center of the IC in order to cause the remaining portion of each of the pins to snap firmly into the slots. A slight rocking motion when pressing the IC may cause easier entry. Removal of the IC from the socket is much less complicated and is done by simply inserting a small screwdriver under one end and gently pulling upward until the IC snaps out. Practice this procedure with a defective integrated circuit if possible because if the pins are badly disfigured, the component may be ruined.

This brief discussion of soldering and building techniques has applied mostly to integrated circuits and other solid-state devices. It should be pointed out that heat damage can occur to any

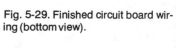
Fig. 5-29. Finished circuit board wiring (bottom view).

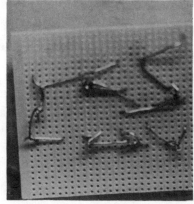

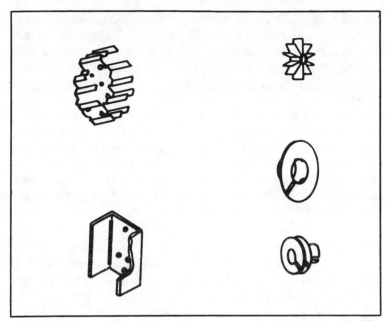

Fig. 5-30. Various types of heat sink devices designed to be mated with transistors or diodes, or both.

electronic component. Small circuit boards generally use small, low wattage components. Resistors, capacitors, and tiny coils can be easily damaged, so solder these connections as rapidly as possible while making sure that you don't rush to the point where cold solder joints are formed.

Large current devices such as SCRs and Triacs are not as heat sensitive and soldering guns may be required to make some of the connections using large diameter wire for high power applications. When power levels increase to levels where the component case cannot dissipate all of the heat to the surrounding air on its own, it becomes necessary to install it on an external heat sink. Heat sinks come in many sizes, with the larger surface areas being able to dissipate more heat. Figure 5-30 shows an assortment of heat sinks which are designed for transistor, SCR, and triac mountings.

Most of the devices which require heat sinks and are purchased new will have the mounting hardware with them. This usually includes a pair of small mica washers, some metal washers, and matching nuts. Also included will probably be a small tube of silicone compound. This is to be applied directly to the device, the heat sink and the washers and other hardware which come in contact with the solid-state device. The silicone is a heat transferrence substance. It allows the conduction of heat away from the component and into the heat sink. It does not conduct electricity. If the silicone is not used, there may be a heat conduction resistance formed between the component body and the heat sink. If this occurs, the solid-state device may be destroyed in a short, operating time.

A heat sink should be mounted so that plenty of natural air is allowed to pass across its surface. In very high power applications, a small cooling fan is used to force air across the cooling fans. For most applications in this book, this should not be necessary, but remember, heat is a destroyer of the efficiency of electronic devices. The cooler you can keep your circuit components, the more efficiently your system will operate.

SUMMARY

Solid-state components are extremely simple devices which may be used to construct rugged, reliable remote control systems. If strict attention is given to their selection, ratings, and mounting, the circuits that are built from them should last a lifetime. Take the time needed to properly design your circuit and practice proper soldering techniques. Generally, you should make your circuits as simple and uncluttered as possible. This will add to the reliability and subtract from the repair time should a malfunction in the circuitry occur. Take the time to make good solder connections but not so long that the delicate device is damaged from the heat. After soldering is completed, allow adequate time for the devices to cool to room temperature before applying power. Many good solid-state devices have been unnecessarily destroyed by applying operating voltage and current a few seconds before they had cooled from recent soldering.

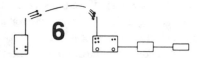

6

Electronic Power Supplies

In this age of inexpensive electronic devices, some of which would have been impossible only a few short years ago, the AC power supply for DC or battery powered devices has become a very affordable and useful piece of equipment. Almost every device which can be powered by dry cell or rechargeable batteries can also find its operating requirements in an AC to DC power supply that is designed to be operated from standard 115-volt house current.

TRANSFORMING VOLTAGES

The internal circuitry of these power supplies first changes the 115 volts from the line into another value through a transformer. This latter value is often 9, 12, or 24 volts. The transformer converts the voltage value. In this case, the value is lowered, thus the transformer is said to have "stepped down" the primary voltage of 115 volts AC. The latter figure, although lower than house current, is still AC (alternating current), but battery-powered devices require DC (direct current). The conversion of alternating current to direct current is easily accomplished by inserting a diode or rectifier at a point in the circuit following the voltage transformation (see Fig. 6-1). Alternating current flows first in one direction and then in the other. Direct current flows in only one direction, thus one side of a DC supply is always positive (+) in relation to the other side which is negative (−). AC current is also positive on one side and negative on the other but only for a fraction of a second. Then, the current reverses and the side which was positive is now negative, and vice versa. This change takes place about 60 times per second in the United States which is described by saying our power is supplied at a frequency of 60 hertz or cycles per second.

After the output from the transformer has been rectified or changed to direct current by the diode, which allows current to flow in one direction but blocks its flow in the opposite direction, we arrive at DC (direct current). This current, however, is still not usable for battery-powered devices because it is not a steady voltage. It will rise to a peak value and then reduce to a zero value. Depending on the type of rectifier circuit used, it may rise and fall sixty times per second or 120 times per second. This rate is directly dependent upon the frequency of the primary or line voltage. The output from a battery does not do this. If the voltage value of a dry cell battery is supposed to be 1.5 volts, then it stays at about this value at all times.

To correct this rise and fall of voltage (called pulsating DC), it is necessary to insert a capacitor between the two sides of the circuit (+ and −). Figure 6-2 shows the capacitor's place in the circuit. The capacitor has the ability to store power. When the pulsating DC voltage reaches its peak, the capacitor is charged to about the same value, but when the pulsating DC begins to drop to zero, the charge in the capacitor immediately begins to add or fire its stored energy back into the circuit, making up for the drop in the rectified voltage. When the rectified voltage has reached zero, the capacitor has discharged much of its energy, but then the pulsating DC begins to rise again, recharging the capacitor and boosting the DC output which is starting to drop due to the discharging capacitor. It is this hand in hand action which stabilizes the voltage to the desired output level. This simple process takes place many, many times each second but is self-sustaining as long as power is being supplied to the input of the transformer and all components are operational.

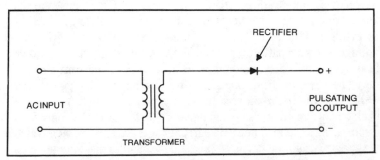

Fig. 6-1. Placement of rectifier in AC-derived power supply for DC output.

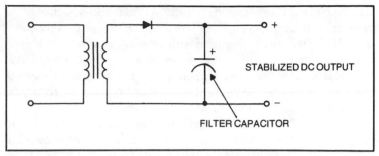

Fig. 6-2. Filter capacitor smooths out AC ripple in pulsating DC output.

FILTERING

When a value of 12 volts, for instance, is rectified and connected to a capacitor as described, this is called filtering. Filtering of this type will affect the 12 volt DC value by increasing its value to as much as 15 volts when no current is being drawn from the circuit. This is the peak value of the pulsating DC voltage. If you wish to power a device which requires a 12 VDC input, this 15-volt value must be decreased. In Fig. 6-3, this is accomplished by inserting a resistor (R1) in series with the DC output. The value of this resistor will depend on the amount of voltage to be dropped or subtracted and by the amount of current the device under power requires. When current is passed through a resistance, a voltage drop occurs. The power which is "used-up" by the resistor is gotten rid of as heat which is radiated from the body of this component.

In this circuit, there is yet another resistor (R2) which is called the *bleeder*. It was stated before that a capacitor connected in the manner shown has the ability to store power. Even after the power supply is turned off or disconnected from the wall outlet, the

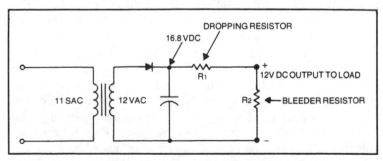

Fig. 6-3. Series dropping resistor inserted in circuit to maintain desired voltage under load.

full value of the power supplied is still stored in the capacitor. The bleeder resistor bleeds off this power which is also dissipated as heat. While we are dealing here with low-voltage power supplies, in a high-voltage circuit, lethal values of power could easily kill a person long after the power supply was completely disconnected from a source of primary power if a bleeder resistor or resistors were not connected to the circuit.

Figure 6-3 shows a very basic and simple power supply. Even so, it will power many types of non-critical battery-powered devices without the need for batteries which run down and must be replaced. These devices would include small hobby experiments which draw the same amount of current at all times.

When more complex devices must be powered from the AC to DC power supply, problems can occur with the circuit shown in Fig. 6-3. Transistor radios, amplifiers, and the like do not usually draw the same amount of current from the power supply at all times. For instance, as a radio is called on to deliver more volume to the speaker, more current must be drawn from the supply. As more current is drawn, the supply voltage drops. When the large current drain decreases, the voltage from the supply increases. Most transistor circuits require a constant voltage at all levels of current demand throughout their operational cycle. As the voltage from the supply fluctuates, the operation of a transistorized device can become erratic. It could even be damaged.

Figure 6-4 shows a power-supply circuit which, while still simple in nature, avoids much of the problem of varying values during operation with devices which draw different amounts of current.

Zener diodes are used mainly for voltage regulation purposes. When the value of a voltage output from a power supply is held relatively constant, the voltage has then been regulated. There are varying degrees of regulation which are determined by how much the voltage value changes over a given current supply range.

The zener diode is always rated at a specific voltage value. When the voltage from the supply is equal to or less than this stated voltage, the diode has no affect on the circuit. But when the power-supply voltage level exceeds the stated value of the diode, the zener conducts and an effective short circuit is placed on the power supply. In other words when the zener value is exceeded, this diode becomes the equivalent of a wire placed between the positive and negative sides of the supply. When a short-circuit exists, additional current is drawn from the supply, but this current

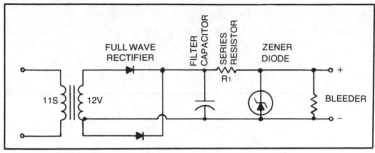

Fig. 6-4. Zener diode regulated supply.

has to pass through the series resistor (R1). We have already learned that when current is passed through a resistor, the voltage level is dropped. Therefore, when the zener conducts current, this additional drain causes the resistor to drop more voltage. This drops the voltage value to a point which is equal to or a bit less than the value of the zener diode and it stops conducting. Each time the voltage rises above the zener diode rating, it conducts again and the process is repeated. Voltage regulation is accomplished by adjusting the value of the series resistor so that the voltage at the zener diode is always a bit more than the printed value. The constant conducting and nonconducting of the zener diode holds the voltage to the stated level of the zener at all times. Again, this process takes place almost constantly and in minute fractions of a second.

If a 12-volt DC output is desired from this type of supply, the zener diode chosen should have a value of 12 volts. The DC output from the capacitor should be higher than this figure to assure that a voltage value of slightly higher than 12 VDC is reached. This value is obtained when using a transformer which converts 115 volts AC to 12 volts AC after it has been rectified and filtered.

The zener diode supply is a vast improvement over the circuit shown earlier regarding voltage regulation, and it will power many types of simple solid-state devices, but it still may not provide the degree of regulation required by frequency sensitive devices and certain types of audio equipment. For powering devices such as transmitters, oscillators, and other equipment of the same nature, a more complex power supply circuit is required.

ELECTRONIC REGULATION

Most commercially available power supplies for solid-state equipment use electronic regulation. While a zener diode supply technically falls into this category, electronic regulation usually refers to the use of transistors in the regulating process. Figure 6-5 shows such a circuit. This supply is typical up to the point where the capacitor is connected to the circuit. After this point, an electronic regulator circuit is employed to keep the voltage at a stable level during periods of varying current drain. This is called a series-pass regulator and uses a transistor which is wired in series with the DC output. A zener diode is also used with this circuit to establish a reference voltage for the transistor to act upon. The value of the zener diode will dictate the DC output, after regulation, from the power supply. Resistor R1 establishes the proper base voltage for the transistor in conjunction with the zener diode.

As in the last circuit, the DC input to the regulator portion of the circuit is maintained at a higher level than the desired output. For a 12-volt supply, this level may be almost double, or 24 volts.

To deliver voltage to the output terminals, transistor Q1 must conduct. Unlike other types of electronic components, transistors can conduct by varying amounts. If the input to the transistor was a value of 24 volts, the transistor would deliver 24 volts to the output terminal if it conducted fully. Most devices are either fully conducting or fully off, but in this circuit, the transistor is allowed to conduct only to the point where it allows a value of 12 VDC to be present at the output terminals. This value is determined by the zener diode at its base. When the current drain from the device under power increases, the transistor conducts to a higher degree which allows this amount of current to be passed without affecting the voltage. When the current drain decreases, the transistor drops its conduction value thus increasing the resistance within the circuit and keeping the voltage level at its established point.

It can be seen that in this circuit the transistor performs like a series resistor with a constantly changing value of resistance. An increased current drain would tend to drop voltage through this imagined resistor, so it reduces its resistance automatically to allow for the increased current. When the current demand is less, the voltage would tend to rise due to less voltage being dropped through heat loss in the resistor, but the transistor increases its resistance so that the current being drawn causes the voltage to drop to or be maintained at a designed point.

Remember, the voltage value in a power supply is directly dependent upon current drain and resistance in the circuit. If the current increases, the voltage will drop unless the resistance is decreased. Likewise, when the current drain decreases, the voltage level will increase unless circuit resistance is increased. The series-pass regulator maintains a constant voltage level by acting as a constantly variable resistor which increases its resistance with decreasing current demand and decreases it when the current demand is increasing. This all takes place constantly and in milliseconds, but the output at the power supply terminals stays constant over widely varying current demands from the device under power.

All of the power-supply circuits used in this basic discussion are the simplest versions of each category, but they can all be used with success in a practical application. As an exercise in practical design of AC to DC power supplies, Fig. 6-6 shows a circuit which can be used to provide a highly regulated output of 9 volts DC at a current demand of about 250 milliamperes, which should be adequate to many types of remote control circuits. By referring to the schematic, it can be seen that this is a simple series pass circuit much like the one just discussed. An additional capacitor has been added for increased stability. This circuit can be put together in a few hours time with a total expenditure of less than ten dollars if you shop around. A parts list is also provided to aid in selection of components.

The circuit receives power from a standard 115 volt AC source and steps this value down to 12.6 volts AC. The rectifier circuit is a full-wave center-tapped design which uses two diodes and the center tap of the power supply transformer. Make certain the diodes are connected correctly in the circuit. A reversal here will prevent the supply from operating. After rectification, the pulsating DC current is fed to the filter circuit made up of a single capacitor. Make certain the positive terminal of the capacitor is connected as shown. The electronic regulator circuit is made up of transistor Q1 and zener diode D3 along with a small, half watt

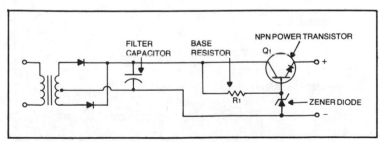

Fig. 6-5. Simple series regulated power supply using npn power transistor.

resistor and another capacitor (C2). A bleeder resistor is not used because the regulator circuit effectively bleeds off stored voltage from the capacitor. Connections of the transistor, zener diode and capacitor C2 must be correct or (inoperation) and possible damage to the various components will result. A 0.5-ampere line fuse is installed at the transformer primary for protection during an overload.

After the circuit is wired, recheck all contacts for proper soldering and to make sure that all components are connected as shown. Insert the plug into the wall socket and connect a voltmeter which will accurately measure 9 VDC across the output terminals. Make certain that the positive lead from the voltmeter is connected to the positive terminal of the power supply. The negative lead is connected to the negative terminal. Throw the switch to the on position. You should get a reading of 9 volts on the meter. This reading may be slightly high or low depending on the accuracy of the voltmeter and the line voltage value in your area. A very high reading or a zero or low reading indicates a problem and the circuit should be checked for the defect or wiring error.

It is most convenient, after proper operation has been established, to wire a standard 9 volt battery connector across the

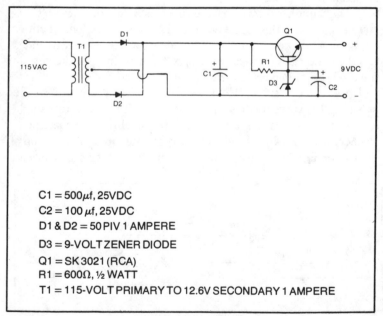

C1 = 500μf, 25VDC
C2 = 100 μf, 25VDC
D1 & D2 = 50 PIV 1 AMPERE
D3 = 9-VOLT ZENER DIODE
Q1 = SK 3021 (RCA)
R1 = 600Ω, ½ WATT
T1 = 115-VOLT PRIMARY TO 12.6V SECONDARY 1 AMPERE

Fig. 6-6. A 9-volt regulated power supply for remote control circuits.

terminals of the supply. Make certain that these connections are made in accordance with correct polarity as a reversal here may damage the equipment you want to operate. Also, be sure the devices you power do not exceed a maximum current demand of about 250 milliamperes. High current drains can damage this supply.

SUMMARY

While this discussion on DC power supplies has been very basic, if you understand the information presented, you can congratulate yourself on the fact that you know more about power supply operation and theory than the great majority of the population. The basics of power supply operation carry through to more complex electronic remote control devices which operate in much the same manner.

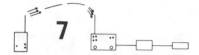

Instrumentation And Units Of Measurement

In dealing with electronic circuits, you will be constantly bombarded by many different units of measurement. Some of these units may even use the same name to describe different electronic values. No matter how far you progress there will always be a need to be familiar with electronic units of measurement; that is, to realize what they mean, and to know how to properly measure them in an accurate manner.

There are many instruments which are used to measure electronic values. Some are multipurpose types which may be able to measure the different values over many ranges. Some instruments are extremely simple and logical to use with electronic circuits in measurement procedures. Others may require more knowledge to enable one to know how to properly connect them to the circuits and also to interpret the indicated readings.

There are several steps to accurate and correct electronic measurements:

☐ You must first know what type of electronic condition you want to measure.

☐ You must know what terms the value of this condition is measured in.

☐ You must know what type or types of instruments are used to measure this particular value.

☐ You must know at what circuit point an accurate reading may be taken.

☐ You must know the proper way to connect the instrument into the circuit.

☐ It is desirable to know the approximate range of the value to be measured so that a proper scale on the measuring instrument may be chosen where applicable.

☐ You must know how to interpret the readings obtained where some form of interpolation or extrapolation is required.

While these steps may sound a bit complicated, you probably are already performing all of these functions when measuring certain types of electronic or electrical conditions. When you measure the voltage of a dry cell battery on a small voltmeter, you are accomplishing all of the conditions for proper measurement as just stated.

☐ You know the type of electronic condition you want to measure—battery potential.

☐ You know that the term of stating this condition is measured in volts.

☐ You know that a voltmeter is needed to measure this particular value.

☐ You know the correct circuit point for taking this reading—the two terminals on the battery.

☐ You know the proper way to connect the instrument to the circuit by placing the voltmeter probes against each contact.

☐ You know the approximate range of the battery voltage to be 1.5 volts for a typical dry cell. You also know how to interpret the readings from the scale on the meter by knowing which range you are in and which scale of the voltmeter to use, or which scale multiplier is used. This is a form of extrapolation when using a multiscale voltmeter.

If you're familiar with the use of a simple voltmeter, then you can see that you're already doing what is necessary to obtain an accurate measurement. This may seem like an oversimplified matter, but measuring the voltage of a dry cell is no more complicated than measuring the frequency of a transmitter, the impedance of an antenna, or the resistance of a complex circuit.

Electronic measurements are sometimes thought of as being very complicated simply because the required knowledge is only vaguely familiar. Before accurate measurements can be taken, it is necessary to learn all you can about various types of electronic values and conditions so that you may fill the required needs regarding the steps mentioned earlier. Every form of electronic measurement will fall into these simple steps and most measurements will be accomplished in an easy, uncomplicated manner as is the measurement of a dry cell battery.

The first step to acquiring this knowledge will be to learn the basic conditions which must be monitored or measured in electronic circuits. We will also study the terms used to describe these values as well as what they mean. It is through this knowledge that the first two steps to electronic measurement are realized. Later, information will be provided that will give you the capability of fulfilling the requirements of the next five steps.

POTENTIAL DIFFERENCE

When measuring the voltage of the dry cell battery, the measurement that is taken expresses a condition of *potential difference*. Referring to Fig. 7-1, we see a typical dry cell battery with a positive and a negative terminal indicated by the appropriate signs. The difference in the two terminals is stated in an electrical term known as the *volt*. The volt is the unit of potential difference and will determine the force of the electron flow in any circuit. A higher voltage will produce a more powerful flow of electrons. The measurement of this flow will be dealt with later, but at the present time we are most concerned with the force created by this potential difference.

The volt can be thought of as being the *basic* unit of measurement of potential difference. Other units which are also used to measure circuit potential differences include the kilovolt which is the equivalent of 1000 volts. Another unit of measurement is the millivolt. This is equivalent to one one-thousandth of a volt. Two millivolts equal 0.002 volts or two one-thousandths of a volt. Another term used to measure electricity is the microvolt which is one one-millionth of a volt. All of these measurements use the term volt, because it is the basic term of measurement of potential force or electromotive force. There is one more term which is rarely used in the measurement of voltage. It describes voltages of a tremendous magnitude. *Megavolt* is a term meaning 1,000,000 volts.

When speaking of voltage, it is essential to know the state or type of voltage in a circuit. Voltage in direct current circuits always maintains a specific polarity. In other words, one pole or contact point is always positive while another is always negative. This is called *DC voltage*. Another type of voltage is constantly reversing its polarity. This is called *AC voltage* and, in house current circuits, it is constantly reversing its polarity 60 times per second.

A meter designed to measure one type of voltage will not measure the other accurately. AC voltage is measured with an *AC voltmeter* while the other type is measured with a *DC voltmeter*.

So, in addition to the other aspects we must know about measuring voltage, we must also know whether it is AC or DC in nature.

Other types of voltage measurements will include radio frequency voltages or *rf voltages*. This is an AC voltage by nature but usually requires a special meter because of the frequency.

There is a third state in circuits which must be taken into consideration when attempting to make certain measurements. This applies only to AC voltages, and is a measurement of the rate at which the electromagnetic field reverses itself or, more simply, how many times the polarity changes from positive to negative and then back again. This is called the *frequency* of the AC voltage.

For most forms of AC, we can assume that the frequency will be on the order of 60 hertz and all common AC voltmeters are designed to operate within this frequency range.

In summary, voltage is a term used to describe *potential difference* or *electromotive force*. The basic unit is the *volt*. There are two basic types of voltage, AC and DC, and a different type of meter is required to measure each voltage. Voltage is the determining factor of the force of electron flow in a circuit.

CURRENT

Another common term used when dealing with electronics describes the amount or quantity of the electron flow in a circuit. This electron flow is generated by the electromagnetic force or voltage applied to the circuit. The basic unit of electrical current flow is the *ampere*. A single ampere is the equivalent of a movement of approximately 6 trillion electrons past a given point in the circuit. This latter definition really has no practical value. Current flow is thought of in terms of amperes and not in terms of how many electrons flow past a point in the circuit. This electron count is given in order to show how the unit of current was arrived at. As was true with the volt, the ampere or *amp* is the basic unit for

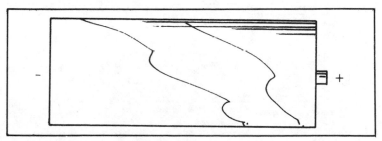

Fig. 7-1. Common dry cell battery showing poles.

other ways of expressing current flow. The *milliampere* is equivalent to one one-thousandth of an ampere and is a term which is quite often used in most conventional electronic circuits. The *microampere*, equal to one one-millionth of an ampere or one one-thousandth of a milliampere, is also used although not to the extent of the milliampere. The term *kiloampere* means one thousand amperes but is almost never used. In most electronic circuits, readings will be taken in microamperes, milliamperes, and occasionally in amperes. High powered circuits will sometimes work in hundredths of amperes.

Current is measured by a device which is basically known as an *ammeter*. The terminology of the meter changes with the different values of current that it is called on to measure. A *milliammeter* measures milliamperes while a *microammeter* measures microamperes.

Unlike the voltmeter, the ammeter must be a part of the circuit in order to obtain a proper measurement of current flow. The voltmeter is placed across the battery contacts with one pole connected to the minus probe or minus side of the meter and the other pole connected to the positive side. Figure 7-2 shows a battery powering a small light bulb with the voltmeter being placed across the battery terminals to measure voltage. Figure 7-3 shows the same circuit with the ammeter placed in such a manner as to measure circuit current. You will notice that the ammeter is a part of the circuit. If the ammeter were removed, the circuit would not

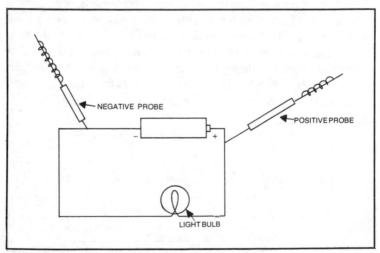

Fig. 7-2. Voltage measurement probe sites in simple circuit.

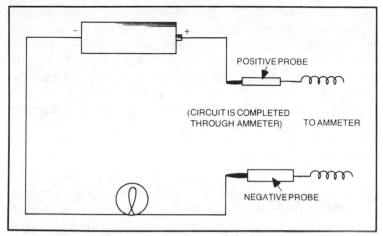

Fig. 7-3. Current measurement probe points in simple circuit.

operate. It will also be noticed that the negative terminal of the ammeter is placed at the positive terminal of the battery. This is due to the fact that the ammeter is placed in series with the circuit. In all series connections, the positive pole of one polarized device is connected to the negative pole of the other. In a parallel circuit, such as the one used for voltage measurement, the voltmeter is placed in parallel with the circuit to be measured and thus the positive terminal of the voltmeter is placed against the positive terminal of the battery. The same is true of the negative terminals.

In the discussion on voltage, we talked about AC volts and DC volts. AC stands for alternating current while DC means direct current. The terms current and voltage are often used together to describe the overall condition of the circuit. As was true with voltage, one meter is required to read direct current and another meter to read alternating current. An alternating current meter is not a polarized device. It does not have a positive and a negative terminal because in an alternating current circuit the positive and negative terminals or polarities are constantly reversing themselves. An alternating current ammeter is still connected into the AC circuit in series, but without regard to any polarity markings. If a DC ammeter were placed in an alternating current, the needle indicator would record one value in one direction and then the needle would reverse itself and go off scale below zero in an attempt to read the current flow in the opposite direction. For this reason, a special alternating current ammeter is necessary to measure AC circuit current values.

While current flow is determined in force by the voltage applied to the circuit, it should be remembered that one ampere of current is *still* one ampere of current regardless of the amount of voltage applied to the circuit. Current coupled with resistance (another circuit value that will be discussed later) is the cause of heating effects in circuits. It is current which causes an electric heater, a stove, and many other devices to become hot when this flow passes through a resistance.

RESISTANCE

When voltage is applied to a circuit, the electromotive force starts electrons in motion for current flow. An electronic circuit is normally composed of various components which affect the flow of current; however, one component, the conductor, is designed to pass current unimpeded to the various other components in the circuit. This conductor usually takes the form of a piece of wire. The term *conductor* is correct when describing this wire because it does conduct the flow of electricity more than it impedes its flow. Any practical conductor, however, also exhibits a resistance to the flow of current. This is known as the resistive properties of the wire, and any other components for that matter will also have a certain resistance to the flow of current. There are three types of materials which make up electronic components; they are conductors, non-conductors, and semiconductors. A conductor, as we have already learned, tends to conduct the flow of electricity while offering a minimum amount of resistance. A non-conductor restricts the flow of current much more so than it conducts it. Just as every conductor does offer some resistance to current flow, every non-conductor will offer some conductivity, but this is a very insignificant factor. A semiconductor material is one which lies between conductors and non-conductors; it tends to pass current about as well as it resists its flow. You may hear the term semiconductor applied to transistors, diodes, and other electronic devices which are made from a type of crystal which has been treated in such a manner as to be classified as a semiconductor. A *resistor* is a device which offers a lumped and specific value of resistance to the flow. This can be thought of as being a non-conductor of a rated value.

The unit of resistance is the *ohm*. The value of one ohm is the amount of resistance which, if connected between the two terminals of a 1-volt battery, will produce a current flow of one ampere. It is easy to see by this definition that all of the terms used

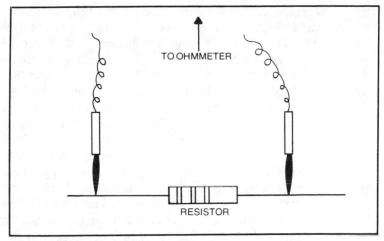

Fig. 7-4. Resistance measurement probe points.

to describe values in an electronic circuit are interrelated and dependent upon one another for proper understanding of definitions. The unit of resistance, the *ohm*, is symbolized with an *omega* sign (Ω).

The resistance of a circuit is measured with an instrument called an ohmmeter. This device operates by passing a small amount of current through a circuit and noting its effects regarding circuit resistance and displaying these effects on the face of a meter. An ohmmeter is usually calibrated in many different scales - from 0 to 100 all the way up to 1 to 10,000,000 ohms in many cases.

The instrument itself will be dealt with later. The ohmmeter is normally placed in series with a circuit to measure total resistance. For individual components, it is placed in an apparent parallel configuration (illustrated in Fig. 7-4) with a resistor. Placing the probes parallel with the component actually produces a circuit which consists of the ohmmeter acting as the power source with each of the probes having a positive and a negative terminal in addition to the component under test being in series with that circuit. An ohmmeter is usually powered by a small battery and applies voltage in such a way as to form a circuit.

An ohmmeter can often be used as a continuity tester by measuring the resistance to current flow in a circuit. Figure 7-5 shows how this is done. The first circuit is composed of a single conductor. When the probes of the ohmmeter are placed at either end of the circuit, the resistance to current flow will be small because a single conductor is used which offers very little electrical resistance. Depending on the length of the conductor, the measured resistance should be less than one ohm or possibly several ohms if the conductor is small in diameter and 10 or more feet long. The second circuit in this figure consists of the same conductor but it has been cut at a central point. This means that there is no direct path through the conductor for current to flow. The resistance of this circuit is said to be infinite, offering an infinite amount of resistance to this flow. Actually, if the severed wires are placed in close proximity to each other, an infinitesimal current flow can occur through the air gap; however, this current is so small due to the resistance being so great that for all practical purposes the ohmic reading is infinity. In that same circuit, if the gap was bridged by placing a 100-ohm resistor in the circuit, then the resistance indicated on the meter would be 100 ohms plus whatever small amount of resistance would be present in the conductor. This is the way most resistance measurements are taken.

ELECTRICAL POWER

The basic unit of electrical power is the watt. In DC circuits, the power in watts which is consumed by a device is equal to the circuit voltage multiplied by the circuit current in amperes. In an AC circuit, the same calculations are performed; however, the product is then multiplied by the power factor of the circuit.

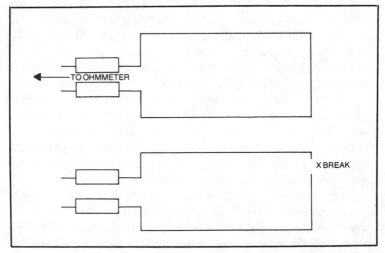

Fig. 7-5. Measuring continuity with ohmmeter. Top circuit is complete. Bottom circuit is open.

Again it can be seen that this unit of measurement is dependent upon the other values within the circuit. Another way of determining power in a circuit is to multiply the square of the circuit current in amperes times the value of resistance in the circuit. It can be seen that the calculations of power in an electronic circuit can be accomplished when only two values are known. These values include; voltage and current, current and resistance, and resistance and voltage.

Power can be measured by using two of the meters we've already talked about. The voltmeter and the ammeter or the ammeter and ohmmeter or even the ohmmeter and the voltmeter can be used when the proper formulas are known to extrapolate or take the information provided in other units and convert it into watts.

Another electronic measuring instrument is called the *wattmeter*. It consists of a combination of the meters described and performs the extrapolation for you, giving the readings in watts on a calibrated scale. This device is normally connected to the input of a circuit or at the output of a transmitting circuit. Figure 7-6 shows a transmitter wattmeter.

Just like the other terms, watt is the *basic* value of electrical power. A *kilowatt* is equivalent to one thousand watts while a *megawatt,* a seldom used term in most remote control applications, is equal to one million watts. The term *milliwatt* describes the unit which is equivalent to one one-thousandth of a watt. Rarest of all in terms of general usage is the *microwatt* which would be equivalent to a millionth of a watt. This latter term is almost never used for electronic applications.

In summary, the watt is the basic unit of power. Two other terms, work and energy, are often used with, or in place of this term. It should be remembered that power, energy, and work are all related but are different terms to describe different circuit conditions or circuit results. Energy is the ability of a circuit to do the work; work is moving something a specific distance; and power rated in watts is the rate at which this work is done.

FREQUENCY

In discussing the measurement of AC voltage and current, the term frequency is used. In the strictest sense, frequency means the rate at which something re-occurs. Frequency is measured in hertz which is a relatively new term and replaces the term "cycle" which

187

was formerly used to refer to the basic unit. Hertz is the name of the man who first studied such electronic conditions.

For electronics work, the term cycle more clearly states just what occurs when we mention frequency. Frequency is most often used to rate an alternating current wave form and describes the rate at which each complete cycle occurs within a specific period of time. The normal amount of time is one second and this combination may be used in describing AC house current which has a frequency of 60 cps which is the same as saying 60 complete AC cycles in one second. The term 60 hertz would be used today which means the same as cycles per second. When we use hertz or cycles per second when describing an AC wave form, we mean that the polarity of the AC current travels from one state to the opposite state and then back again. This is one complete cycle. The AC voltage will travel from a value of 0 to a peak positive value, back to 0 again, to a peak negative value, and then again to 0. This is a complete voltage cycle. Figure 7-7 shows a typical voltage wave form. This shows one complete cycle. In house current, this cycle

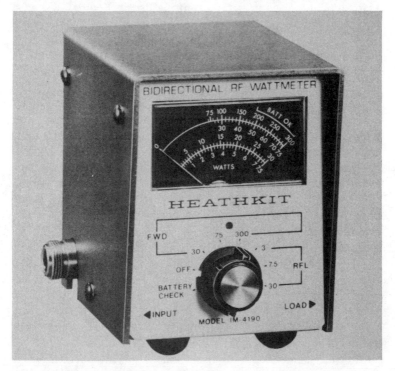

Fig. 7-6. Heath directional wattmeter for power measurements.

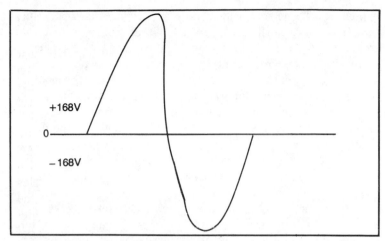

Fig. 7-7. Single cycle of AC sine wave showing zero, peak positive, and peak negative points.

occurs 60 times every second. 60 hertz is said to be the frequency of standard house current.

The measurement of frequency is usually performed on a frequency meter or frequency counter. The frequency counter has replaced the frequency meter in recent years due to the inexpensive availability of solid state components. Frequency counters normally give their readings displayed in digital readout circuits, and the frequency is read directly so it is not necessary to compute the actual figures from a set of readings which was the case with some of the older style meters.

While the hertz is a basic unit of measuring frequency and means the equivalent of cycles per second unless otherwise specified, other derivatives of this term include *kilohertz* which is equal to one thousand hertz; *megahertz,* one million hertz; and *gigahertz* which would be the equivalent of 10,000 megahertz. The hertz is the smallest term used to describe frequency. Any signal which exhibits a frequency of less than one cycle per second is considered to be direct current.

Most frequency counters will measure frequencies starting at just above the direct current range and extending well up past a thousand megahertz. The less expensive models on the market sell for under $100.00 and will measure frequencies up to about 40 megahertz which extends into the very high frequency region of the radio spectrum. Counters to measure higher frequencies than this require more complicated and critical circuitry and are more expensive. These instruments normally include an input which samples a very small portion of the alternating current power source. This is done without actually interfering with the function of the circuit being measured. For example, a small portion of the radio frequency output of a transmitter might be fed to the input of a frequency counter and the frequency measured at all times. The frequency counter then is connected in parallel with the circuit under test. Figure 7-8 shows a frequency counter of the digital readout variety. For AC circuits which contain power outputs in the very low range, a special probe is often used which amplifies the low output to a level which can be used by the frequency counter.

Frequency counters often use high stable quartz crystals to establish an internal frequency rate which is of a known value and a known tolerance. The incoming frequency is combined and compared with this known frequency, and the difference is displayed on the digital readout panel as the actual frequency. Other types of frequency counters use the frequency of the AC line from which they are powered as a time base or reference.

CAPACITANCE

A capacitor is a device which stores electrical energy. In electronics, capacitors are often used as filters in DC power supply circuits to smooth out the pulsating DC which is obtained after the alternating current has been rectified. Capacitors are used also to pass alternating current signals while completely blocking direct current. In many electronic circuits, DC and AC are present in the same circuit. A capacitor may be connected to a circuit to allow the

Fig. 7-8. Heath frequency counter with digital readout indications.

AC signals to pass on to other portions of the wiring while keeping the direct current blocked within its own portion of the circuit.

The basic unit of capacitance is the *farad*. Unlike other units described, the farad is a unit of such great quantity that is not really possible to accomplish in electronic circuits. A capacitor which would be equivalent to one farad in rating would be huge by any comparison. More conventional capacitors will be rated in *microfarads* which equal one one-millionth of a farad. Another rating is the *picofarad* and was formerly known as the micro-microfarad. The picofarad is equivalent to one one-millionth of a microfarad. Capacitors used in DC power supplies to filter the AC ripple component are usually rated in microfarads, while capacitors used for bypassing and feed-through purposes in electronic circuits are often rated in picofarads.

Capacitance can be measured in several ways. A capacitance meter may be used which compares unknown capacitors to capacitors within the instrument of a known value. The difference is indicated on a meter. Another way of rating a capacitor is by means of a grid-dip oscillator. The use of this instrument in measuring capacitance will be discussed later. The measurement of capacitance in electronic circuits is not normally done. Capacitors for specific electronic purposes are purchased from the manufacturer at the values desired. The needed circuit capacitance can be determined from various mathematical formulas, and a commercially made component with the ratings needed is purchased and wired into the circuit.

Measuring capacitors with a capacitance meter or grid dip meter requires extreme caution when filter capacitors rated in microfarads are used. As was previously stated, capacitors have the ability to store electrical energy. In a high voltage power supply circuit, the capacitors can retain the full output of the power supply if they are not discharged through a *bleeder* resistor. In other words, a filter capacitor could be removed from a 4000 volt power circuit, placed on the test bench, and if it were not properly discharged before removal, a lethal potential of 4000 volts would still be present at the terminals for days or even weeks. Theoretically, the charge would stay there forever until discharged, but a slow drain of energy is present from outside the component such as moisture in the air, etc. Even though this capacitor was completely disconnected from the rest of the electronic circuit, it would still have the full potential of the DC output of the power supply. Coming into contact with these terminals would be exactly equal to placing your body across the 4000-volt output of the power supply. Many individuals have been killed by disregarding the energy storage capabilities of capacitors. Whenever they are to be measured in high voltage circuits, make certain that they are completely discharged by placing the shaft of an insulated screwdriver across the terminals. This shorts the circuit within the capacitor and drains it of all stored energy should the regular protective circuits within the power supply fail.

INDUCTANCE

An inductor is a device which usually consists of a coiled wire and offers considerable opposition to alternating currents which are changing polarity at a rate for which the coil was designed. Inductance is measured in a basic unit called a *henry*. The henry, like the farad, is a very, very large value, so more practical units in electronic circuits are usually rated in *microhenries* which are equivalent to one one-millionth of a henry or in *millihenries*, one one-thousandth of a henry.

Coils or inductors are often measured with the grid-dip oscillator. In order to do this, the coil must be connected with a capacitor to form an LC circuit. This will resonate at a specific frequency. When the value of the capacitor is known and the frequency of the coiled capacitor combination is known, then the inductance may be found by dividing the square root of the capacitance value into 0.159. With this form of measurement, no direct reading of inductance is obtained. The grid-dip oscillator will provide only enough information for you to be able to mathematically compute the inductance. Capacitance is also measured in a similar manner. Here, the value of the inductor must be known.

A grid-dip oscillator is pictured in Fig. 7-9. It is actually a small, low-power oscillator whose frequency is controlled by a plug-in coil. This coil is closely coupled to the circuit formed by the inductor and the capacitor connected in parallel. Figure 7-10 shows how the device is coupled by placing it very near the coil. Once the oscillator is coupled to the circuit, the tuning capacitor of the oscillator is adjusted while observing a meter which is placed in the oscillator circuit. When the oscillator is tuned to a frequency of the LC circuit, the meter will dip or drop to a minimum value and then begin to rise again. The oscillator capacitor is adjusted in the other direction until the indicator needle on the meter drops back to its minimum value. When this is accomplished, the oscillator scale is read which is calibrated in hertz. This will give an indication of the frequency of the LC circuit.

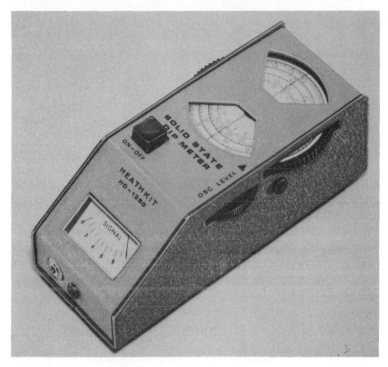

Fig. 7-9. Heath grid-dip oscillator.

The grid-dip oscillator provides only a very rough indication of frequency and therefore only an approximate value of inductance or capacitance. When measuring unknown values of inductance or capacitance, extremely accurate readings are not normally necessary. The grid-dip oscillator is another instrument used to measure frequency, but this is resonant frequency which is an indication of what frequency a particular circuit will produce when under power and not a measure of the actual output frequency.

While the values and conditions covered in this chapter are basic, they are used every day in electronic measurement. Even the most complex procedures and devices will be measuring in some form or other the basic values or a combination of the values discussed in this chapter. Some electronic values will contain capacitance, inductance, resistance, and even more of the basic values we have studied here. It is absolutely necessary to be familiar with the material that has been provided thus far. All of the discussion in future portions of this text will revolve around the facts, material, and ideas presented. If there is any portion of the material that confuses you in any way, go back and reread the section in question. It will save you considerable time in our future discussion and you will absorb more of the information which is yet to be provided.

THE MULTIMETER

One of the most useful and necessary instruments for making electronic measurements is the *multimeter*. This instrument is used more than any other and a day rarely goes by that its use is not dictated when experimenting with or servicing electronic equipment. The multimeter, Fig. 7-11, offers a tremendous range of measurement scales. It will also measure many different values in electronic circuits.

The multimeter is often referred to as the *ohmmeter*, the *voltmeter*, the *ammeter*, or the *dB meter*. All of these terms are technically correct because even the most inexpensive models can usually perform measurements in all of these categories. A multimeter is usually self-contained, supplying its own power source; but sometimes it will require 110 volt house current. Some models are called vacuum tube voltmeters and use vacuum tubes to provide very high accuracy in voltage measurements. Most of these, however, have long been replaced by solid-state equivalents which provide even better accuracy.

Some multimeters are available with a standard meter readout of the values being measured, while others are now available in digital readout forms which provide easy-to-read digital displays of electronic values. A multimeter will usually have two test probes

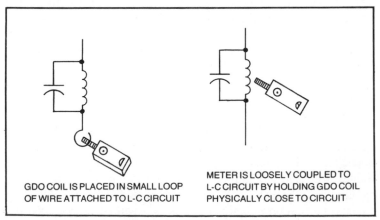

GDO COIL IS PLACED IN SMALL LOOP OF WIRE ATTACHED TO L-C CIRCUIT

METER IS LOOSELY COUPLED TO L-C CIRCUIT BY HOLDING GDO COIL PHYSICALLY CLOSE TO CIRCUIT

Fig. 7-10. Correct connection procedures for using grid-dip oscillator.

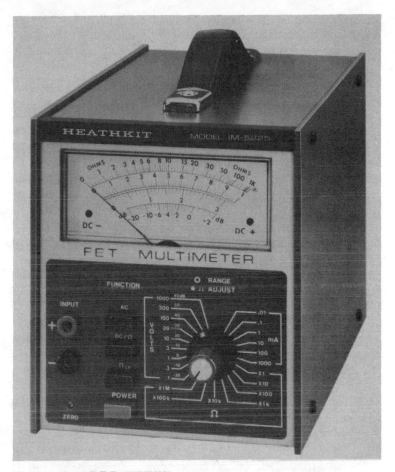

Fig. 7-11. Heath F.E.T. multimeter.

that are used to measure values such as voltage and resistance in
addition to amperage, without inserting them in different sockets.
A mode switch on the multimeter automatically switches the probe
leads into the correct portion of the multimeter circuit. Some of the
more expensive multimeters will contain three probes. One is a
ground connection when measuring voltage circuits, while the
other is the positive probe for DC circuits. When measuring AC
circuits, the DC probe contains a switch which is thrown to the AC
position and the same probe is used for the latter circuit. The
ground lead is also used in conjunction with the third lead to
measure resistance. In this mode, the voltage probe is not used.
Sometimes for high voltage or high current measurements, a

separate probe slot is provided. In this case the positive probe is removed from its normal slot and inserted into the other in order to obtain the special readings. Most multimeters measure from 0 to several million ohms and from 0 to about a thousand volts AC or DC with a special probe slot for up to about 5000 volts.

Figure 7-12 shows a multimeter which provides several different scales on one meter face plate. One scale is used to measure ohms and is usually calibrated from 0 to infinity. When the resistance range switch is in the times 1 position, the meter will be read according to the printed scale. In the times 10 position, all readings are multiplied by 10. The same is true of the voltage scale which is used to measure voltages in several different ranges by simply multiplying by the factor indicated on the value switch.

The multimeter is a very accurate device when a good instrument is purchased. The accuracy is determined by the quality and tolerance of the internal resistors used. A high quality instrument will have a good majority of its price applied to the precision resistors. An inexpensive unit will contain resistors which are not as close to stated value due to a larger tolerance of error. These latter instruments will give generally good measurements, but they can vary as much as 10 to 15 percent, and in some cases even 20 percent from the actual value being measured.

Almost every electronics shop or test bench will contain a multimeter in some form or other. When a voltage is present in a circuit and its value needs to be known, it is the multimeter which is used. The same is true when a resistor needs to be checked, a battery is thought to be weak, or the current drain of a circuit needs to be known. We've already learned how the various meters are connected to the electronic component in order to obtain an accurate measurement. The multimeter probes are placed into the electronic circuit in many different ways depending on the electronic condition being read. For instance, if DC voltage is being read, the probes will be placed with the negative one at circuit ground and the positive probe on the positive pole of the circuit. The selector has already been thrown into the DC voltage mode and a range selected which should be somewhere in the area of the anticipated voltage. If the ballpark value of the unknown voltage cannot be determined, the multimeter range switch is set to measure the highest DC voltage possible. When the probes are properly placed and the circuit activated, the range switch will then be adjusted downward until a proper reading is indicated. This is a safety precaution and protects the instrument from damage. If the

Fig. 7-12. Heath Multipurpose VOM.

probes of the multimeter are being used to measure a circuit which contains a potential of 500 volts DC and the range switch is in the 100-volt position, as soon as the circuit is activated the indicator will go completely off scale. That is, past the 100-volt position, which is the maximum it is designed to read with the mode switch at this setting. If a standard meter indicator is used, the meter needle may be severely bent. The internal mechanism of the meter may be damaged and some of the precision resistors within the multimeter circuit may be destroyed. This is why it is necessary to have a rough idea of the range of values you are measuring. If you are in doubt, set the range switch to the highest value available and then work downward until the range approaches the value you are measuring. This method applies only to measuring voltage or current with the multimeter. When measuring ohms, if the meter goes off scale, no serious damage will result.

Most multimeters have one or two adjustments on them to properly set the meter range so that precise measurements can be taken. It is often necessary to *zero* the meter. This means that when the probes are shorted together, the meter scale will read zero. This adjustment may be found on the meter face itself in the form of a tiny set screw, or the adjustment may be made by a *zero adjust* potentiometer on the multimeter panel. When measuring resistance, it is also necessary to set the scale for the high end. This is done by using the *ohms-adjust* control. To properly set this control, it is necessary to first zero the meter with the mode switch in the ohms position and in the appropriate range. With the two probes shorted together, the *zero adjust* is turned until the meter reads zero. This control is often touchy and some very fine adjusting may be necessary. Now separate the test probes and the needle should travel toward the far end of the scale. The last reading on the ohms scale will be infinity, often labelled *inf*. This is an indication of infinite resistance and is what is obtained when the test probes are separated from each other having only open space in between. By adjusting *ohms-adjust* control, the indicator is made to read exactly at the *infinity* marker. Adjusting this control will often have an effect on your meter's zero adjustment, and you may have to re-short the test probes and adjust for a zero reading again. It will then be necessary to go back and touch up the *infinity-adjust* control. Each control has an effect on the other, but after a few short adjustments, the scale should be calibrated so that the needle indicator reads *zero* when the probes are shorted, and *infinity* when they are separated. For truly accurate measurements of resistance, this alignment process must be done each time you switch to the ohms position from another point on the mode switch. When moving from one ohms scale to another, this procedure will also have to be repeated.

Never apply the probes of the multimeter to a power circuit under power when the mode switch is in the *ohms* position. This can result in damage to the meter. When the mode switch is in the current measuring position, be extremely careful of how you place the probes within the circuit. If the probes are placed in parallel with a power circuit, the meter will act as a short circuit. This would probably cause severe damage to the multimeter and might even damage delicate components within the circuit under test. It was learned earlier that for current measurements, the ammeter must be placed in series with the circuit under test. This is true when using the multimeter, which in the current measuring position, is an ammeter or to be more accurate a milliammeter and a microammeter as well as an ammeter in some cases. The black probe from the multimeter is the negative probe and is connected to the positive point in the circuit which has been broken to allow for insertion of the ammeter. The red probe is the positive one and is placed across the other side of the open contact point to complete the circuit for proper current measurements.

As before, if you are not certain of the ballpark range of current values within a circuit being tested, set the current scale to the highest position with the value selector switch. When the circuit is activated, if no reading is shown, adjust the range switch to a lower value until the proper scale is found. The most accurate measurements will be obtained with the multimeter and most other meters when a range is chosen which places the indicator in the upper half of the meter scale. This applies only to conventional pointer-type meter indicators. Digital readout meters will normally provide accurate readings over their entire range.

In summary, a multimeter will measure all practical values of resistance, voltage, current, and decibels. This latter measurement is actually a determination of sound level and is the equivalent of a measurement of AC voltage. To get these measurements, however, the test probes have to be connected into the circuit in the correct manner, (parallel or in series) and you must know the proper setting at which to place the multimeter mode switch. Most of this knowledge comes from an understanding of the values you wish to measure and their relationships in the circuit.

Resistance Measurement

It has already been learned that resistance is measured in ohms and indicates the resistance to the flow of current in an electronic circuit. In order to accurately measure resistance, we must first gain an understanding of how resistances can combine in a circuit to produce different or unusual values.

Figure 7-13 shows two resistors valued at 100 ohms each in a parallel circuit. When the probes of an ohmmeter are placed across each of these resistors as single units and not connected as shown, the meter will indicate 100 ohms. But when resistors are connected in parallel, the total resistance of the parallel circuit will be less than the resistance of any one component. A practical formula for determining resistance in parallel is also shown in Fig. 7-13. Substituting the values for R with the actual values used in the circuit of 100 ohms, the formula works out to 100 × 100 which is equal to 10,000 divided by 100 + 100 or 200 for a total of 50 ohms. The actual value is a fraction of an ohm less than this formula indicates, but it is as accurate as most applications require.

Resistances in parallel offer two paths for electric current to flow. This is the reason that this combination offers less resistance than a single resistor of the value of any one of the resistors combined in parallel. When measuring resistors that are combined in the circuit with other resistors, it must be remembered that some of the other components will often be wired in a parallel configuration with the resistor under test and will affect the readings you obtain. It may not be possible to measure the true value of a single resistor until it is disconnected from the circuit by clipping or desoldering one of its leads. This effectively removes the resistor from the rest of the electronic circuit, and it may be measured as a discrete component.

Figure 7-14 shows three resistors in a parallel combination and, again, the same formula can be used by substituting the values of R1 and R2. The answer from this formula is then substituted

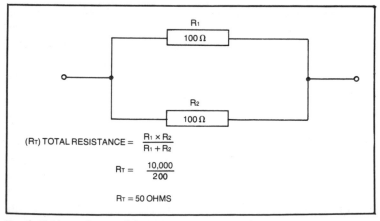

(R_T) TOTAL RESISTANCE $= \dfrac{R_1 \times R_2}{R_1 + R_2}$

$R_T = \dfrac{10,000}{200}$

$R_T = 50$ OHMS

Fig. 7-13. Measuring parallel resistors.

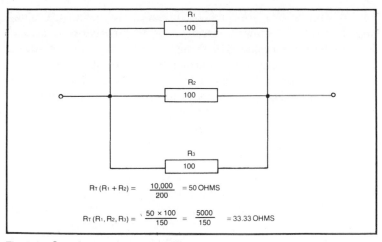

$$R_T (R_1 + R_2) = \frac{10{,}000}{200} = 50 \text{ OHMS}$$

$$R_T (R_1, R_2, R_3) = \frac{50 \times 100}{150} = \frac{5000}{150} = 33.33 \text{ OHMS}$$

Fig. 7-14. Complex parallel resistor circuit.

back into the formula again for the value of R1 and the remaining resistor value is substituted for R2. When the formula is worked again using these values, the total resistance of the parallel circuit which utilizes three resistors will be known.

Figure 7-15 shows the proper placement of the multimeter probes to measure a circuit composed of parallel resistors. This is a very easy circuit to measure as the probes are placed across only a single resistor; one probe on one lead and one probe on the other. While the probes are shown attached to only one resistor, the leads of any other resistor in the parallel circuit may be used. The indicated reading will remain the same.

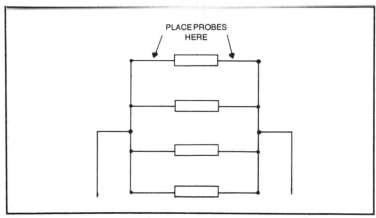

Fig. 7-15. Measurement probe placement points for complex circuit.

When resistors are combined in *series*, the resistance values *add*. Figure 7-16 shows a circuit using four resistors in series. Each resistor has a value of 10 ohms; therefore, the total resistance is 40 ohms. The formula for determining total resistance in a series circuit such as the one shown is: Total Resistance = R1 + R2 + R3 . . . . This formula applies for the number of resistors in the series circuit. The same figure also shows the placement for measuring a single resistor in the series string. You may place the probes in such a manner as to measure any number of resistors in a string without removing them from the remainder of the circuit as was necessary in a parallel configuration. While this does not pertain to multimeter measurement, resistors which have stated power values will add those values when combined in the circuit in either a series or a parallel configuration. Ten 1-watt resistors will have a total power rating of 10 watts when combined in a series or a parallel circuit. The power rating of both circuits would be identical. The resistance value of the two circuits would be different, however.

Figure 7-17 shows resistors combined in a *series-parallel* circuit. This is a more complex configuration and uses two separate circuits; one consisting of two resistors in series. The two circuits are then connected in a *series-parallel* combination. It is a very simple task to figure the total resistance of this circuit by treating each of the two circuits as separate entities. Determine the total resistance in each circuit and then determine whether the two circuits have been combined in a series or parallel configuration.

This circuit can have its total resistance measured quickly and accurately through the use of a multimeter. This precludes the necessity of performing the mathematics involved to determine total resistance from the individual resistor values. Figure 7-18 shows the correct placement of the ohmmeter probes for determination of total circuit resistance. The actual measurement will match very closely the total resistance found from the formula.

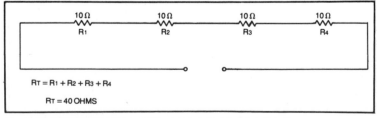

Fig. 7-16. Determining total resistance in series resistor circuit.

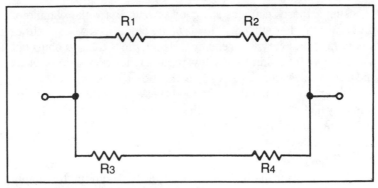

Fig. 7-17. Series-parallel circuit.

Other components in an electronic circuit can affect the accurate measuring of resistance when only the resistance of a certain portion of that circuit is desired to be known. A transformer connected across a resistor will cause the resistance to read near zero regardless of the value of the component. A solid state diode or rectifier will cause the resistance to appear very low when the ohmmeter probes are placed across the component leads in one way. The reading will appear closer to the actual value of the resistor if the ohmmeter probes are reversed, however. When you desire to know the value of a single component or of a single circuit within the framework of a more complex circuit, it may be necessary to isolate the circuit under measurement.

In the measurement of the resistance of solid-state rectifiers or diodes, an ohmmeter can be a very effective device, generally

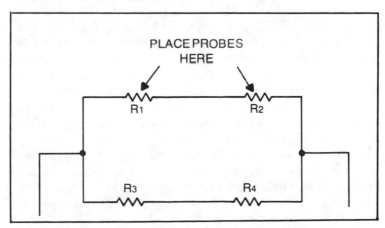

Fig. 7-18. Probe placement points for complex circuit components.

checking if these rectifiers are good or bad. Place the ohmmeter probes across the component leads of the diode. You will get either a very high or a very low reading. (Usually less than 10 ohms but sometimes a bit higher for the low region.) In reversing your black and red probes from one lead to the other, you will obtain an opposite reading if the diode is good (very low if the first reading was very high or very high if the first reading was very low). This is just another use of a very versatile instrument.

DC Voltage Measurement

The use of the multimeter as an accurate DC voltmeter simply requires the changing of the mode switch to the proper position for DC voltage measurement. The black lead or probe is the negative or ground connection, whereas the colored lead or probe is the positive connection. It is desirable to know where the circuit ground connection is for placement of the negative lead (if a negative ground is used). Many pieces of electronic equipment which are constructed on a metal chassis will use the chassis as the ground tiepoint. By placing the negative probe across the chassis, the positive probe may be brought into contact with various voltage points for accurate readings. Some electronic circuits have a positive ground. This means that all other voltage points are negative with respect to the chassis. Normally, the chassis will serve as the negative connection point. Figure 7-19 shows a complex circuit which is powered by a 9-volt battery. Here the negative probe of the ohmmeter is placed across the negative terminal of the battery. The voltmeter scale will be set so that 9 volts will not exceed its range of measurement (the 10 - or 15-volt DC scale will most likely be used and is usually found on most multimeters). It may be necessary to decrease the scale range down to 5 volts or so as various points of the circuit are probed because some of the voltages will be dropped in value by the various resistors. But it is reasonable to assume that no potential which is higher than the battery voltage will be encountered if this is the only source of power. Now that the negative probe and the meter range adjustments have been set, it is a simple matter to probe various parts of the circuit for an indication of voltage. A DC voltmeter exhibits a very high resistance between its two probes and thus has very little effect on the circuit which it is connected to. A warning was stated earlier in this text about the placement of ammeters in the circuit because of their very low internal resistance. An ammeter connected across the negative and

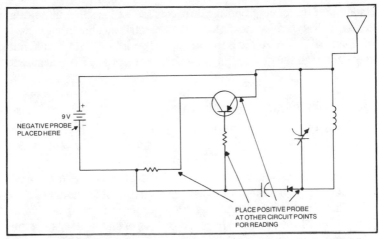

Fig. 7-19. Measuring various voltage points in circuit by placing negative voltage probe at negative terminal of battery supply.

positive terminals of a battery will create a short circuit because the two points are connected by the equivalent of a very low resistance conductor. A voltmeter has a very high internal resistance and is the equivalent of a very large value resistor being connected across the same two terminals. Only a tiny amount of current will flow. In a voltmeter circuit this current is sampled and converted to a voltage reading on the meter scale. It is easy to see that it is not possible in most circuits to cause circuit damage from taking voltage measurements. This is provided that you don't accidentally short circuit some components by having the metal tip of the probe accidentally touch two circuits at the same time.

Another note of caution: When measuring medium and high voltage circuits, always connect the negative probe to its negative source so that you don't have to hold that probe in your hand. Now, when using the positive probe use whichever hand you're comfortable in working with and place the empty hand in your pocket. This could prevent serious injury or death. If your left hand should be resting on the metal chassis (which is circuit ground) and your right hand with the positive probe in it should accidentally come into contact with a medium to high voltage source inside the equipment cabinet, the full potential could pass through your heart by traveling through your arm and chest to complete the circuit. A severe electrical shock can often cause the heart muscle to become temporarily paralyzed. The heart stops beating. If someone is not present who can perform external heart massage, you, the victim could die. By keeping one hand in your pocket at all times when taking voltage measurements, any electrical shock you may receive will be limited to only one side of your body and should not travel through the heart muscle. The chances of a serious accident are greatly reduced.

Capacitance and Inductance Measurement

Two electronic values which have been briefly discussed earlier in this text, capacitance and inductance, are often used simultaneously in circuits to arrive at even more complex readings of impedance. As was previously stated, capacitance is measured in farads while inductance is measured in basic units of henrys.

The purpose for discussing both of these units in the same chapter is that, often, to arrive at the values of one, the value of the other must be known. This applies especially in transmitting circuits or other types of L-C connections which determine frequency in remote control filters. A value of known inductance coupled with a value of known capacitance will result in a circuit which has a specific *resonance*, another term used to describe tuned circuits.

The measurement of individual units of capacitance and inductance has already been discussed, so in this portion of the text, we will be learning the proper methods to measure these combinations of capacitors and inductors.

THE GRID-DIP OSCILLATOR

Again, the grid-dip oscillator comes into play as a most convenient method of determining resonant frequency. The grid-dip oscillator is an electronic device which can be basically described as a low-powered transmitter or oscillator with a large frequency range which is tunable. Additionally, some provision has been included to indicate grid current or base current for solid-state grid-dip meters. This instrument normally has an external coil which allows the oscillator to be tuned over a great range. One coil may be designed for a small segment of the high-frequency spectrum while another will take over for the portion of the spectrum where the first leaves off.

The grid-dip oscillator is coupled to the circuit under test by placing its coil close to the circuit. The meter of the oscillator is then observed while the tuning control is moved slowly through its range. As the frequency of the grid-dip oscillator approaches the resonant frequency of the L-C circuit, the meter indicator will decrease. At a point where the oscillator frequency matches the frequency of the circuit, the meter will neither increase nor decrease its reading but will remain in a valley called a *dip*. Further adjustment of the oscillator tuning control will cause the meter to rise.

When the dip is reached, the operator knows that the grid-dip oscillator is producing a frequency which is equivalent to the resonant frequency of the L-C circuit. He then determines what the frequency of the grid-dip oscillator is by:

☐ Reading the calibrated scale on the face of the grid-dip oscillator which indicates the output frequency.

☐ Tuning a calibrated receiver in the general range of the grid-dip oscillator frequency until the oscillator's output tone is heard. Tuning of the receiver is done with the bfo in the on position. The receiver is tuned until the monitored signal decreases in frequency to a point where it vanishes from the monitor speaker or headphone. This is called the *zero beat* of the signal. When this is accomplished, the frequency of the grid-dip oscillator can be determined by the frequency reading of the more accurately calibrated receiver.

The method for determining the frequency of the grid-dip oscillator as described in step two is to be preferred over the method described in the first step. This is due to the fact that grid-dip oscillators are not highly stable or accurate devices. The closeness of nearby metal objects can slightly alter the frequency of the oscillator output and cause the calibrated scale to be substantially incorrect for determining frequency. A calibrated receiver on the other hand acts directly upon the grid-dip oscillator signal and is a much more accurate form of determining frequency.

It should be remembered at all times that the grid-dip oscillator is, at best, only a rough indicator of resonant frequency. The basic instability of this device is necessary for it to operate properly. The calibrated face of the grid-dip oscillator can be used to indicate the frequency of its output signal, but the method previously described of monitoring the signal on a calibrated receiver will provide much better accuracy where this is required. Even this latter method is still not highly accurate. Stray capacitance and inductance is almost always involved in the measurement of resonant circuits and will affect the accuracy of the indicated readings. Also, the amount of coupling of the grid-dip oscillator to the circuit will alter the readings substantially. Coupling is the process of holding the coil of the grid-dip oscillator in close proximity to the resonant circuit. This allows the capacitive and inductive components of the resonant circuit to be electronically "coupled" to the circuitry of the grid-dip oscillator.

THE IMPEDANCE BRIDGE

Though normally used in only the more demanding measurements, the impedance bridge has a definite use in modern measurement techniques. *Impedance* is the total opposition offered by a circuit to the flow of alternating current of a particular frequency. Impedance is a combination of ohmic resistance as well as capacitive and inductive *reactance*. Impedance is measured in ohms. Although this is the same unit of common resistance, the measurement of impedance is different from that of pure resistance.

While a simple ohmmeter will measure DC resistance (the resistance of a circuit to the flow of direct current), it will not measure impedance (the resistance to the flow of alternating current in a circuit). To measure circuits for specific values of impedance, it is necessary to have a testing instrument which can produce a flow of alternating current within the circuit under test. The impedance bridge (Fig. 7-20) does just this. While not all impedance bridges contain their own alternating current sources, all of them are designed to operate with an external generator as a source of alternating current.

An impedance bridge provides extremely accurate measurement of resistance, capacitance, and inductance. It will also measure the dissipation factor of capacitors and the storage factor of inductors. While actual use of various types of manufacturer's bridges varies, most of them require that the operator monitor the alternating current which is in the form of an audio signal on a meter or in a set of headphones while adjusting the various controls until a complete "null" is obtained. This null is the cancelling of the audio signal. When the signal is no longer heard, the calibrated controls are consulted and the various readings obtained. Impedance bridges can be rather demanding as far as a delicate touch is concerned in tuning for a null. It may be necessary to adjust the various controls repeatedly as one will usually have an effect on the other.

The majority of electronic measurements which are required to be made by the hobbyist or the general experimenter will not require the use of an impedance bridge. However, certain critical applications may demand its use.

Fig. 7-20. Heath impedance bridge for measuring complex values of resistance and reactance.

LINE FREQUENCY

The frequency of AC current in the United States is maintained at 60 Hz. Although this value may vary by one or two Hz in some areas, the value is usually constant. European line frequency is usually maintained at 50 Hz, but most equipment designed to operate at 60 Hz will operate as well at 50 Hz, and vice versa.

It is not usually necessary to measure the frequency of AC voltage, but this can be done by using a frequency counter which is designed to read values at this level. The frequency meter is normally coupled to the AC line through capacitors which are designed to sample only a small portion of the AC potential. Different types of frequency counters will have different instructions for measuring the AC frequency.

Some sources of electric current will have a line frequency of 400 Hz. These are rare but are encountered from time to time. 400 Hz energy was popular on aircraft during World War II and many types of surplus equipment are still designed to be operated from current of this frequency. A circuit which is designed to operate from 400 Hz line current will not operate from current of a 60 Hz value. As a matter of fact, the 400 Hz circuit will become hot and will probably begin to burn if adequate fusing has not been provided. This frequency rate may also be measured by using a frequency counter.

For most practical applications, the rate of frequency of the AC line will be of little consideration because of its standard value of 60 Hz. If this value is not proper for circuit conditions, there is

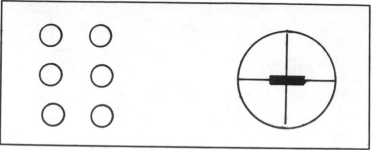

Fig. 7-21. Oscilloscope display with input to vertical amplifier.

nothing that can be done to directly alter the frequency rate of the line, and a source of alternating current of a different frequency must be derived by electronic means.

THE OSCILLOSCOPE

Most of the instruments discussed so far in this chapter have provided indications of electronic conditions by displaying equivalent readings on a meter. Indeed, many of the instruments used to measure electronic values have the word "meter" in their designated names. The voltmeter, ammeter, ohmmeter, etc. all use meters to indicate circuit conditions. Meters are normally designed to respond only to a specific electronic value; therefore, when metering an electronic circuit, we get only a very restricted indication of what is taking place. For most electronic measurement purposes, the limited range of meters is no problem. But to monitor other types of conditions, this limitation becomes a severely restricting factor, and another more versatile means of measurement is needed.

The *oscilloscope* is a device which actually shows a picture equivalent to the electronic conditions in a circuit being tested or measured. When you watch your television, which is a type of oscilloscope, you are seeing an electronic equivalent of the signals which are being fed to its circuits. These signals are of course the transmissions from the various television stations.

For most electronic measurements, the oscilloscope will display a far less complex picture than that which is received on your television. Basically, most oscilloscopes have two main inputs to which signals from circuits under test are connected. The first input is called the vertical axis, and a signal fed to this input will produce an indication on the cathode ray tube of the scope similar to the one which is shown in Fig. 7-21.

The second input to the oscilloscope is called the horizontal axis and it produces a "picture" similar to the one shown in Fig. 7-22 when a signal is fed at this point. The larger the value of the signal is at the inputs, the longer or wider will be the line or beam deflection on the face of the tube.

The oscilloscope will normally be used with both inputs connected to the electronic circuit under test. Each axis has an effect on the other. Figure 7-23 shows the equivalent picture when equal signals are fed to the vertical and horizontal inputs simultaneously.

Basic Oscilloscope Function

The heart of the oscilloscope is the cathode ray tube (CRT) which focuses an electron beam on the screen which shows up as a single dot when there is no signal input. This focused beam of electrons is then altered by two sets of deflection plates. One set is the vertical deflection plate, the second is the horizontal.

The vertical amplifier increases the input signal which would normally be of a very low value for electronic measuring purposes. This amplification increases the deflection sensitivity of the cathode ray tube. Most oscilloscopes offer a variable control to adjust the amount of amplification. Very low level signals will require the highest amplification, while the high level signals will require little. Some scopes even offer an attenuator which is a simple circuit that decreases the input of the signal to be measured. Most cathode ray tubes will exhibit a deflection of the focused beam of one inch for every 125 volts of applied signal to the deflection plates. However, with the addition of this amplifier, the sensitivity is increased tremendously so that a signal of less than 1/10th of a volt may produce the same amount of deflection.

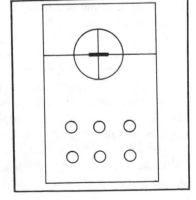

Fig. 7-22. CRT display when horizontal amplifier is driven.

In the same manner as the vertical amplifier, a horizontal amplifier is provided to provide input to the horizontal deflection plates. In most instances, the gain of this amplifier is much less than that of the vertical amplifier. It does, however, work on the same principle and in the same manner as the vertical amplifier.

The next basic section of the oscilloscope comprises the *time-base* circuitry which drives the horizontal amplifier. A time-base is simply a wave form generator whose output is in the form of a sawtooth wave. It is then a simple oscillator with a distorted output. The main purpose of the time-base generator is to provide horizontal deflection of the electron beams. A sawtooth waveform is made up of a voltage that rises evenly with time. This is called a linear function. When coupled to the rest of the oscilloscope circuitry, it causes the electron beam to move across the CRT screen at a constant and linear speed. This time-base section is sometimes referred to as the *clock* because it allows the beam to move through each division on the rectangularly ruled graticule which divides the face of the cathode ray tube (CRT) into equal segments.

The oscilloscope will show an accurate picture of the signal which is fed to its inputs. Sometime, the vertical input only will be used and the time-base circuit will internally drive the horizontal amplifier. At other times, the time-base generator will be electronically removed from the circuitry and an external horizontal input used. Due to the ability of the oscilloscope to accurately display waveforms, when a 100 hertz sine wave signal is fed to the vertical input of the oscilloscope, a sine wave pattern is displayed on the CRT screen. The timing circuit is normally calibrated and a signal of unknown frequency can be accurately measured by proper alignment of a single wave on the scope graticule. When this is accomplished, one need only read the settings of the time-base controls and a fairly accurate determination can be made of the frequency. The more expensive scopes offer better calibration, and thus better measurement accuracy.

Most service oscilloscopes which are manufactured today use vertical circuits that are known as video amplifiers. These circuits are identical to the video amplifiers in television receivers. They are also known as wide band amplifiers. These scopes will amplify frequencies of as high as about 5 megahertz. Less expensive scopes may have vertical amplifier bandwidths of as little as 100 kilohertz (kHz). These are usually limited to audio frequency applications. Some very expensive oscilloscopes have special amplifying sections which are good to 30 megahertz and higher. Scopes with this bandwidth comprise some of the most expensive pieces of test equipment manufactured today.

The type of instruments discussed so far have had vertical amplifiers which will accept the input from AC signals only. Oscilloscopes, in addition to being classified as narrow band and wide band, are also classified as AC coupled or DC Coupled. An AC scope utilizes an AC coupled vertical amplifier whereas a DC scope has its vertical amplifier DC coupled. A DC oscilloscope can also be used with AC input, but it offers the advantage of being able to accept an input of 0 Hertz for pure DC. Most AC only scopes will not accept an input of less than about 60 Hertz. In many applications, it is necessary to measure signals of a lower frequency. Here, a DC scope is an absolute necessity.

Oscilloscope Probes

An oscilloscope probe is a device which couples the signal from the equipment under test to the vertical amplifier of the scope. A probe can be as simple as a single piece of wire which connects the equipment under test directly to the vertical amplifier. In its most complex form, the probe will consist of a separate electronic circuit which may offer signal processing functions instead of merely coupling to the vertical amplifier.

A main purpose in using an oscilloscope probe is to match impedances between the test circuit and the vertical amplifier input. Most oscilloscope vertical amplifiers have a very high input impedance, usually in excess of 1 megohm shunted by about 25 picofarads of capacitance. A high impedance is ideal for testing purposes because it avoids the possibility of a short circuit. Problems can develop in circuits, however, which will not supply sufficient signals to a high impedance source.

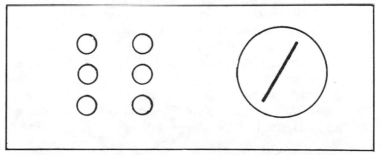

Fig. 7-23. Resulting CRT waveform when vertical and horizontal amplifiers are equally driven.

Another type of probe is called a demodulator. This can be thought of as a type of signal processing probe which contains a diode detector. This probe is used to extend the effective high frequency range of a scope into the very high frequency region. It does this by demodulating the input signal from the modulation envelope and passing the pulsating DC onto the vertical input. While the signal it is demodulating may be many times higher than the effective frequency response of the oscilloscope, the probe passes along only the audio information contained in the modulation waveform of the signal. Therefore, most service oscilloscopes can adequately respond.

Using the Oscilloscope

The oscilloscope is used for monitoring, testing, and measuring almost every type of electronic function. The scope will give more information, more accurately, than almost any other type of electronic instrument. The oscilloscope is not necessarily an expensive piece of equipment, although the overall price will be higher than most single measuring devices. It must be thought of as a combination measuring instrument; one which can perform a myriad of tests and measurements while retaining peak accuracy. Mechanical meters are limited in their response to electric pulses and conditions which occur within a fraction of a second. They are limited because a mechanical movement must take place. However, the oscilloscope has no mechanically moving parts and can respond to the instantaneous amplitude of all input signals. It is for this reason that the oscilloscope is often the pivotal test instrument in electronic shops and laboratories throughout the world.

TRANSISTOR CHECKER

There are many parameters, conditions, and readings connected with transistors. It is impossible to measure all of these by using simple instruments. However, it is possible to determine many of them by using a single electronic unit called the *transistor checker* (Fig. 7-24). Several of these instruments are available from different manufacturers. Some perform only a few tests while others perform many, simultaneously. Some of the better models do not require the connection of coded instrument leads to specific transistor component leads. In other words, any of the three leads emanating from the transistor checker may be connected to any of the three leads of a bipolar transistor. A portable transistor checker is shown in Fig. 7-25.

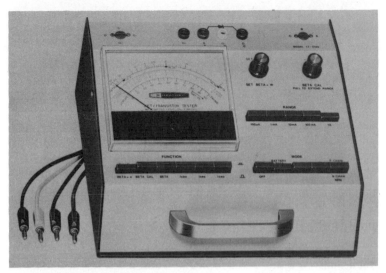

Fig. 7-24. Heath electronic transistor checker.

The internal circuitry of the transistor checker is not of concern to us in this discussion on electronic measurement. Only the indications given by this instrument will be considered important.

A good transistor checker will determine the state of bipolar transistors, field-effect transistors, or silicon controlled rectifiers out of the circuit or even in the circuit. It will also identify good and bad diodes in or out of the circuit. The transistor checkers which do not require that their test probes be connected to specific component leads will identify the emitter-base-collector leads, or the gate lead when testing field-effect transistors. This is the basic testing that is performed on the solid state components. Some checkers offer even more readings such as an indication of whether the semiconductor material is silicon or germanium, or the basic makeup of the layering of semiconductor material (pnp or npn; n or p channel for field-effect transistor testing).

This may seem like a complicated process and it is, but all of the complication has been worked out within the internal circuitry of the instrument. Normally, all that is required of the operator is the connection of the transistor to the probes and the rotating of a multi-switch through about six positions.

Using a checker, an unknown transistor can be tested, and it can be determined whether or not the component is good or bad and, if good, it can identify the type of material the component is made of, what type of component it is, its polarity, and its lead configuration. All of this information can be obtained in less than 10 seconds.

Transistor checkers are in wide use for most types of electronic testing and measurement today. The instruments are relatively inexpensive, costing anywhere from $15 to about $250 for standard, commercial units. Even the least expensive model is far superior to a multimeter for testing the performance of solid state devices. The ohmmeter is still occasionally used for tests on transistors, but only in emergency situations where a transistor checker is unavailable. For checking integrated circuits, a more complex instrument is often required. For these applications, the IC tester in Fig. 7-26 is often used.

BIPOLAR TRANSISTOR RATINGS

Earlier, the various terms for describing certain electronic conditions were discussed in detail. Modified terms are used to describe conditions and ratings of solid state devices. All of these use terms which have already been discussed to describe them, but it is important to know exactly what these conditions mean when used to describe operating conditions of solid state devices.

Most solid-state devices are described in two ways: limiting conditions and characteristics. Additionally, the manufacturer will usually give each device a manufacturer's type number and a packaging type. This refers to the case the component is mounted in.

Certain symbols will be used as abbreviations for describing characteristics and operating parameters of these solid state devices. While there are many such abbreviations, only the most prevalent ones will be discussed here. These are the abbreviations which you will see most often and which are most important when considering remote-control operation.

The first abbreviation stands for device dissipation. It is abbreviated P_t. This rating is most often expressed in watts, although in certain devices it may be expressed in milliwatts. P_t describes the maximum amount of power in watts that can be dissipated by the component and still remain within the manufacturer's ratings.

Another important rating is that of collector current, abbreviated I_c. I_c describes the largest amount of current which can be drawn through the collector of the transistor without exceeding manufacturer's ratings. This condition is normally described in amperes of DC current although it may be described in milliamperes for lower power devices.

The third condition used to describe bipolar transistors is breakdown voltage and is expressed in three different forms. First is the collector to base breakdown voltage which is abbreviated V_{cvo}. The second abbreviation is V_{ceo} which is a description of the collector to emitter breakdown voltage. The third and final abbreviation describes the emitter to base breakdown voltage and is displayed as V_{ebo}. Breakdown voltage is the maximum applied voltage to the various portions of the transistor that can be tolerated by the device. When these ratings are exceeded, the

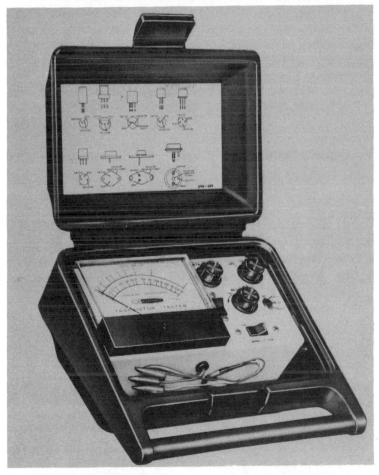

Fig. 7-25. Portable transistor checker.

device is being operated outside of manufacturer's ratings and may soon fail to operate.

The terms just discussed are used to describe limiting conditions on solid state devices; in this case, bipolar transistors. These are the maximum manufacturer's ratings and should not be exceeded.

In addition to limiting conditions, most solid state devices will also be described as to operating characteristics. One of these characteristics is abbreviated h_{FE} and stands for the typical current gain of the device. This is also known as the amplification factor. This figure is not stated in any specific term but uses a single number to indicate the multiplication factor. A second characteristic is the typical gain bandwidth, abbreviated f_T. Generally, this indicates the maximum frequency the transistor will operate at while still exhibiting typical characteristics. The typical gain bandwidth may be given in kilohertz (kHz) or in megahertz (MHz).

Finally, most manufacturers will list the type of case or cases in which the solid state device is housed. Certain manufacturers may offer the same device in several different cases. Case style and electronic lead configuration or basing will usually be described with numbers such as TO-1, to TO-92, etc.

RECTIFIER RATINGS

Solid-state diodes or rectifiers already discussed have their own set of ratings which differ from those of bipolar transistors. These ratings are also abbreviated and it is important to understand these and the symbols when measuring the devices. The two main limiting conditions which need to be known concerning rectifiers and diodes are *peak reverse voltage*, abbreviated PRV and the *forward current*. The peak reverse voltage is the maximum voltage that a diode will withstand when it is not conducting or when it is blocking the flow of current. This condition is expressed in volts and sometimes in kilovolts. The maximum forward current rating, which is abbreviated I_f, indicates the amount of average current which may be drawn through the device while still staying within the manufacturer's ratings. Two other ratings are the *nonrepetitive voltage* and the *surge current*. The nonrepetitive voltage is expressed as V_{rm} and describes the maximum amount of voltage the device can withstand on a one time or nonrepeating basis. This may be a bit confusing at first, but it should be understood that in certain electronic circuits, outside line current conditions can cause a phenomenon known as a voltage "spike" to occur. This spike may be 10 times higher than the normal line voltage, and while occurring for only a minute fraction of a second, can place a serious strain on solid state components. The V_{rm} rating indicates the device's ability to withstand this type of condition.

The surge current rating, abbreviated I_{fm}, is expressed in amperes of DC current and is similar to the nonrepetitive voltage rating in that it does not describe a continuous operating condition. When electronic devices are initially activated or turned on, there is a tremendous surge of current through many of the circuits for a small fraction of a second while the filter capacitors in the power supply are charging. Also, devices which use electric motors require a tremendous amount of current to get the motor started. Once the shaft is spinning, a much smaller amount of current is required to maintain operation. Surge current can be over 200 times the average current in a circuit, and therefore surge current ratings are often many times the average current or I_f rating of a rectifier or diode. When diodes are used in applications which constitute high surge current demands, this rating becomes extremely important for device survival.

ZENER DIODE RATINGS

A zener diode is a special type of solid-state device which is used for voltage regulation purposes. The device is so designed

Fig. 7-26. Heath integrated circuit tester.

that it will not conduct until the applied voltage has exceeded a specific, predetermined value. Zener diodes are therefore available in many different regulator voltage values. Regulator voltage is abbreviated V_z and is normally expressed in DC volts. If the output from a zener regulated power supply should be 12 volts, then the zener diode for use in this type of supply would have a V_z rating of 12 volts DC. The device is inserted in the power supply circuit in such a manner that all voltage values above 12 volts DC are conducted to ground by the zener. This, in turn, creates a voltage drop across a series resistor in the power supply circuit and effectively drops the voltage back to the 12-volt value. Similarly, if the desired output from the power supply were to be 22.5 volts, a 22.5-volt zener diode would be chosen.

Another limiting condition of zener diodes is expressed as I_z and describes the maximum regulator current that may be drawn through the device. This rating is normally expressed in milliamperes (mA) and occasionally in amperes (A).

A third limiting condition is that of device dissipation. This describes the maximum amount of power the zener diode may dissipate without exceeding manufacturer's ratings. This is known as the P_d rating and is described in watts or milliwatts.

SUMMARY

When dealing with the electronic aspects of remote control, it is imperative to know the basic theory behind the various devices and components which will be encountered. It is equally important to know how to use basic measuring tools and instruments and the correct methods of making the measurements. Only through electronic measurement can the circuits be certified as working properly. Even if the proper remote control functions are being sent, received, and acted upon in a normal manner, a quick check of the operating parameters with the correct instrument will make certain that none of the components are operating outside of their ratings. This condition can be tolerated for a short period, but the system will eventually fail as the components are destroyed. These failures often occur at the most inopportune times.

Once your circuits are completed, it is imperative that tests and measurements be made to determine the exact operating conditions of the various components. After this has been accomplished, the project can finally be termed completed.

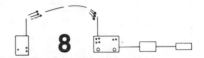

8

Sound-Activated Control Systems

A relatively simple way to remotely control a system is through sound activation. This is most applicable for systems where the remote control point is physically close to the work load. In a sound-controlled system, a predetermined tone or other controlled sound is generated at the remote point. A circuit which is placed at the work load receives this sound and activates a relay or other type of switching device which controls the system.

The sound can be anything from a pure, undistorted tone to a clap of the hands. This will depend on the environment at the work load and near the remote control point. For more advanced and versatile control functions, a number of tones may be used to key more than one function. This will mean that the tone receiver and processor will have to discriminate against tones of different frequencies, selecting and routing the proper one through a filter.

Figure 8-1 shows a block diagram of a single channel system using a microphone pickup to receive the tone and provide an output to a circuit which is, basically, a sound activated switch. When sound is picked up by the microphone, the circuit receives it in the form of a very low current. This is amplified and fed to the switch which may be a normally-open or normally-closed relay, depending on the switching function desired.

Figure 8-2 shows the actual switching circuit which uses an SCR to rectify and to control the switching of the main power source. This can be of any type, although, here, a small transformer with a 115-volt primary and 12-volt secondary is used to supply voltage and current to a 12 volt DC relay. This circuit is the heart of the switch with the audio amplifier and microphone

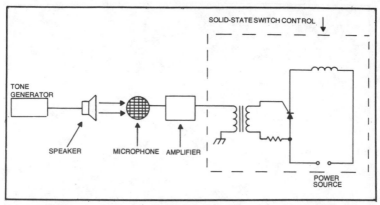

Fig. 8-1. Single-channel tone activation system using a microphone pickup.

supplying the command signal from the remote sound source. The relay is a latching type, so one sound or tone will be required for an instant to activate the switch. A second tone causes the relay to open again. The small, audio transformer matches the output of the audio amplifier to the SCR circuit. Triggering is controlled by the variable resistor.

This can be a very effective remote-control circuit but it does not discriminate from various sounds. Microphones can be chosen which have a very limited audio range, but any sound of sufficient volume which lies within the mircrophone response range will trigger the switch. This is often undesirable, and accidental triggering can occur from unwanted sounds. For this reason, this circuit is not highly reliable, although it can be used to turn small appliances on and off as long as the control point is somewhere near the work load.

Figure 8-3 shows a tone-activated control which has a simple discriminator circuit located between the pre-amplifier and the amplifier to pass only a very narrow range of audio tones. Tones which lie outside of the filter range are greatly suppressed, and proper amplifier input and output adjustments along with the right sensitivity setting of the variable control will allow very stable switching using a single, fixed tone. Accidental triggering will very rarely occur with this circuit.

The discriminator circuit is placed in series with the connection between the microphone pre-amplifier and the input to the amplifier. The circuit is simple and can usually be made from stock components in a short period of time. It consists of a coil and a capacitor wired in parallel with each other and in series with the

222

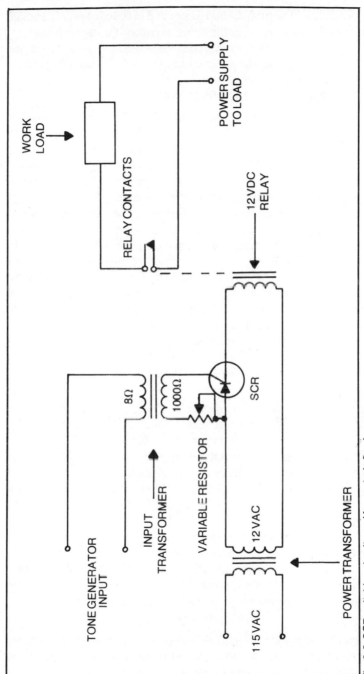

Fig. 8-2. SCR switching circuit used for control of main power source.

audio circuit. The preamplifier is needed in order to supply enough drive to the amplifier after the signal is passed through the filter.

Filter design frequency determines which tone or tones are allowed to pass from the preamplifier through the amplifier and on to the SCR switching circuit.

FILTER CALCULATIONS AND CONSTRUCTION

Filter frequencies are determined by the relationship of the coil or inductor to the value of the parallel capacitance. Several combinations may be had, all passing the same tones. For instance, if a filter is designed to pass a tone at a frequency of 1000 hertz, decreasing the value of either the inductor or the capacitor would tend to increase the frequency of the passband of the filter. But, if the inductor value were increased by a set amount and the capacitance were decreased by a similar percentage, the tuned frequency of the circuit would not change. For this reason it is possible to use many different types of inductors which are available on the commercial and hobby markets to build tone filters. All one needs do is connect the correct amount of capacitance in parallel with the inductor as derived from the basic formula for determining the frequency of an L-C circuit such as the filter being described. This formula reads:

$$f = \frac{1}{2\pi LC}$$

This means that the filter frequency (f) is equal to one divided by the product of two times pi (3.14) times the square root of the inductance times the circuit capacitance (LC). These latter values must be inserted into this equation in their basic units. This means henrys for the coil and farads for the capacitor. These are horribly large terms and inductors are usually rated in thousandths of a henry or smaller while capacitors used for filter design will most often be rated in millionths of a farad or smaller. It takes a calculator which will provide readings in scientific notation to conveniently work these calculations for tone filter circuits.

A more convenient formula for working the filter frequency can be arrived at by converting the inductance value to microhenrys and the capacitance to picofarads. Then, the formula reads:

$$f = 1,000,000/2\pi LC$$

A good calculator may still be needed, but at least the figures involved do not require 14 place readouts. The frequency will be expressed in kilohertz using this latter formula.

Construction of these types of filters is totally non-critical and simply involves the wiring of a capacitor in parallel with the inductor. The filter can be installed in a separate, aluminum box or mounted on the board where the other parts of the circuit are located if this type of construction is being done. It is a good idea to place the filter as closely connected to the preamplifier as possible. The amplifier, in turn, should be closely connected to the output of the filter. This prevents stray pickup of noise and static.

It is most difficult to have to resort to the complicated formulae already discussed in this chapter each time you want to build a noncritical tone filter. This can be easily overcome by buying a variable inductor. Several types are available at moderate costs which may be adjusted from about 50 millihenrys to over 200 millihenrys. This makes an ideal filter arrangement, because the inductor may be adjusted to select different tones to be passed. An inductor of the approximate, adjustable values discussed can be wired in parallel with a small capacitor of 0.15 microfarad value. This combination will provide a variable passband for audio tones from 500 hertz to over 1700 hertz. If you need a wider range of tones, extra capacitors can be switched into the circuit. This latter arrangement is shown in Fig. 8-4.

Of course, one L-C circuit will pass only one tone or narrow range of tones at one time. An adjustable component may allow for resetting the pass frequency whenever desired, but only one tone for one setting can be passed at a time. This is not a hindrance for a single channel remote control system which only requires a single command for activation and another to cease operation, but the more complex systems will need something else to establish multi-channel command control.

Figure 8-5 shows a block diagram of a system which allows three-channel operation. It is basically the same circuit as before,

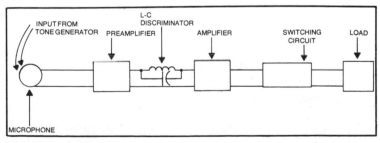

Fig. 8-3. Block diagram of tone control system showing placement of discriminator circuit.

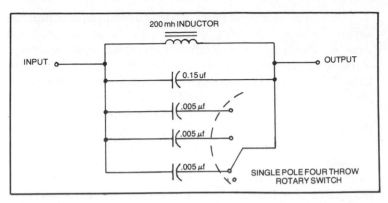

Fig. 8-4. A rotary switch can be used to connect additional capacitance into discriminator circuit.

but the L-C filter circuit is placed at the output of the amplifier instead of the output of the preamplifier. Each L-C output is connected to a separate SCR switching circuit. Each filter is designed to pass a separate tone. If filter "A" will pass tones in the 1000 hertz range, then a 1000 hertz signal at the audio pick-up will pass through the amplifier and provide activation current for the "A" circuit only. The other two circuits will reject the 1000 hertz tone, as they are designed to pass two, other frequencies. If circuit "B" is set up with a frequency pass of 1800 hertz, when this latter tone is sensed at the pick-up, it will be passed on to the "B" circuit while circuits "A" and "C" reject the tone.

An interesting point of this circuit is realized when two tones are simultaneously fed to the pick-up. If dual tones of 1000 and 1800 hertz are induced at the input, then both circuit, A and B will be triggered. All three circuits can be activated simultaneously by a three tone input at the designated filter frequencies.

This system can be expanded by adding more filters and SCR circuits without ever changing the preamplifier and amplifier electronics. There may be some matching problems with a multicircuit hookup between the output of the amplifier and the input to the SCR circuits, but generally, this is not a severe problem in most systems. The variable resistors in the SCR switches may just have to be set for maximum sensitivity.

ULTRASONIC CONTROL

The previous circuits used *audible* tones for control. This is satisfactory for non-critical applications, but audio tones are generated through many sources such as music, human voices, and normal household appliances. If the pick-up system is activated, accidental system triggering can result. A better system would use the *ultrasonic* band of frequencies which lie outside of the human hearing range. These systems are often used with television receivers which offer remote channel switching control.

Figure 8-6 shows a Heathkit color television which offers an excellent remote control system. Such functions as "on-off", channel change, volume control, etc. are all offered to the user from a convenient hand-held control. Referring to Fig. 8-7, when a control function is sent by depressing a button on the ultrasonic transmitter, it is received by the ultrasonic transducer (or pick-up) and sent to the proper integrated circuit amplifier which sets on the input to drive and control the various circuits in the television receiver.

Many of the ultrasonic transistors are mechanical in nature rather than electronic. They generate a burst of ultrasonic tone by mechanically striking a metal bar within the transmitter case. The transducer produces an electric signal when an ultrasonic tone is received.

Stepping switches or relays and electric motors connected to the various controls within the receiver perform the work. For instance, when the ultrasonic transducer receives a tone command to increase volume, its electric signal is sent to the volume amplifier which sends pulses of amplified signal to a stepping switch ganged to the volume control. As the switch moves through its positions, the volume would be increased accordingly. An alternate command to decrease volume, sent by the ultrasonic transmitter, is handled in a like manner, but a different amplifier is

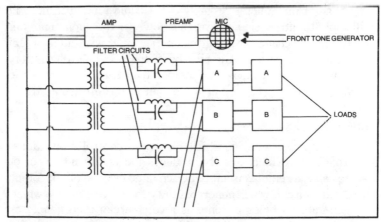

Fig. 8-5. Three-channel tone-activated system.

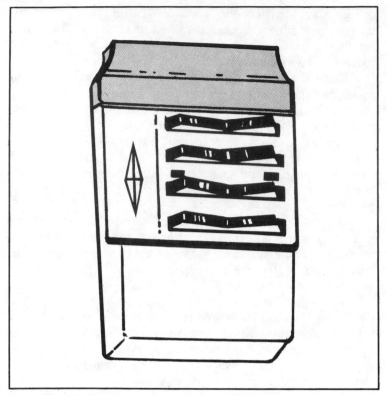

Fig. 8-6. Heath ultrasonic function control.

used which reverses the stepping switch and drops the volume level.

Most of the general circuitry of the audio tone systems will still apply to ultrasonic applications, however the audio amplifier will have to be changed to an ultrasonic receiver with audio output to directly trigger the SCR circuits of the previous discussion. Many integrated circuits will offer direct triggering action from the electrical signals produced at the output of the ultrasonic transducer.

TONE ENCODERS

An encoder is a device which, for remote control applications, electronically or mechanically produces the command signal or *code* to cause a remote circuit to be activated or controlled. A tone encoder is a circuit which generates audio tones. One device which will generate all types of tones is a commercial tone generator shown in Fig. 8-8. Normally used for testing response in stereos

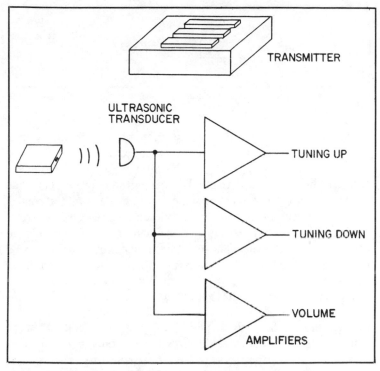

Fig. 8-7. Diagram of basic functions of ultrasonic system.

and other audio systems, the generator may be directly attached to the filter-amplifier circuitry or may be connected to an audio amplifier and speaker for a wireless system. This latter design is shown in Fig. 8-9. The low-level output from the tone generator is

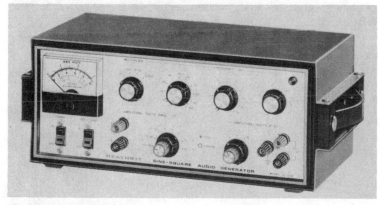

Fig. 8-8. Heath audio generator.

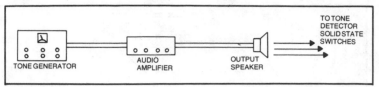

Fig. 8-9. Attachment of commercial tone generator to audio amplifier and output speakers.

amplified and broadcast through a standard speaker to the awaiting transducer at the workload. The output from the transducer is routed to the amplifier, through the filter network, and, finally, triggers the switching circuit. The generator offers a very wide range of audio tones, most of which will not be used for control functions. In practical application, five tones or less are generally all that one needed. Fortunately, tone generators are very simple to build and cost little. These latter, homebuilt circuits can be custom designed to offer only the tones needed, and their switching circuits are preferable to those of the commercial generator for remote control applications.

Figure 8-10 shows a schematic diagram of a tone encoder which offers a total of five audio tones in the 1200 to 3500 hertz range. This is a basic twin-T circuit using only two, common transistors, three capacitors and a few fixed resistors. The variable resistors are used for accurately setting the frequency of the output tones. These are selected by pushbutton, momentary switches which produce a tone at the output only when the button is depressed. This is a single oscillator, so only one tone may be transmitted at a time. A simple audio amplifier at the output of this encoder will allow the tones to be broadcast across a room to control a work load. The receiving system is the same as the SCR systems discussed earlier.

The circuit should be built on a piece of perforated vector board which can be purchased at most hobby stores. Wiring is totally noncritical. The finished circuit should be mounted in an aluminum box and the switching and frequency control components mounted through the case. The audio output is high impedance with a peak to peak voltage of about 7.4 volts when using a nine volt transistor radio battery. This battery should provide long hours of service, because current is drawn only when a tone is activated. At other times, it is completely off. This encoder was designed for use with latching systems which require only a short tone burst for activation or control. If used in this manner, the battery should last almost its normal shelf-life. If a lot of tone switching and heavy

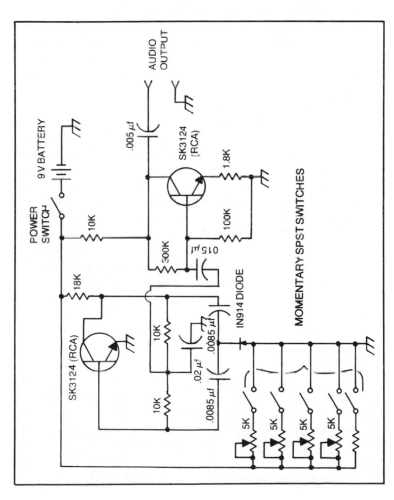

Fig. 8-10. Tone encoder offering five adjustable tones in the 1200 - 3500 Hz range.

usage is anticipated, it might be good to use two batteries in a parallel arrangement to provide nine volts at double the amp-hour rating. This will result in longer battery life.

Figure 8-11 shows a schematic of a tone encoder which uses an integrated circuit and very few other components to generate tones in the 500 to 2000 hertz range. The output is approximately 2000 ohms impedance which may directly drive some audio amplifiers without having to use a matching transformer. A regulated six volt power source is used to supply operating current, although this may be replaced with a nine volt battery by adding a 390 ohm, half-watt resistor between its positive terminal and the rest of the circuit. This will make the generator portable and will not tie the operator to a house current line.

The frequency is controlled by adjusting the variable resistor which is a low wattage potentiometer of 10K ohm value. A similar switching arrangement may be added along with other potentiometers to be able to select several, fixed tones. The previous tone encoder circuit may be used to add this modification.

This circuit can be completed in a short time, and none of the parts are especially critical nor is the wiring. The circuit is mounted on a small piece of perforated circuit board and soldered from the board bottom. The IC leads and those of the other components are simply inserted through the circuit board holes. The connections are then wrapped and soldered, leaving the top of the board very compact and tidy. Audio amplifier circuits also contained in this chapter may be built on the same piece of circuit board for a totally self-contained unit, although two, separate power sources may be required. Both can be obtained from batteries for a portable encoder or derived through transformation and rectification of the AC line current for a fixed design.

Using ICs in remote control circuits often reduces the size of the completed control unit. ICs may also reduce the complexity and wiring which would be present if separate components were used.

AUDIO AMPLIFIERS

The tone encoders discussed in this chapter may often be applied directly to the switching circuits through transformers if a direct-line remote control circuit is desired. However, if you want to broadcast the tone across an area, it will be necessary to increase the audio output through amplification. It is not usually necessary to have a very powerful output, so small, low-current devices can often be used to build the amplifier circuits. Miniature

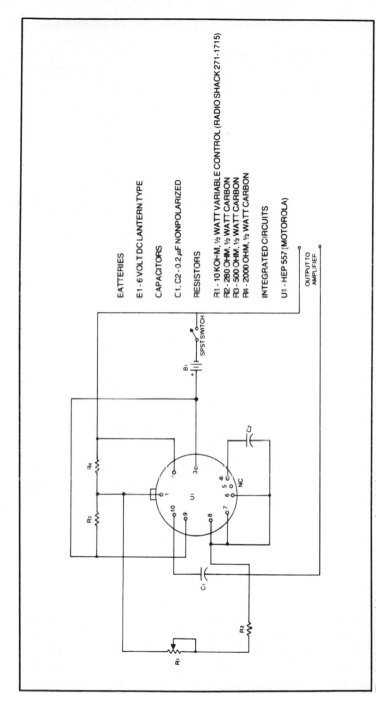

EATTERIES

E1 - 6 VOLT DC LANTERN TYPE

CAPACITORS

C1, C2 - 0.2 μF NONPOLARIZED

RESISTORS

R1 - 10 KOH-M, ½ WATT VARIABLE CONTROL (RADIO SHACK 271-1715)
R2 - 280 OHM, ½ WATT CARBON
R3 - 500 OHM, ½ WATT CARBON
R4 - 2000 OHM, ½ WATT CARBON

INTEGRATED CIRCUITS

U1 - HEP 557 (MOTOROLA)

OUTPUT TO
AMPLIFIER

Fig. 8-11. Integrated circuit tone encoder with frequency range to 2000 Hz.

speakers serve as "antennas" for these audio broadcasting systems.

Figure 8-12 shows a typical circuit which uses an IC as the heart of the amplifier. A miniature gain control is also included to adjust the audio output to the lowest level usable with the decoder at the switching circuit. The circuit draws very little power and operates into a small 8-ohm speaker. It produces a fraction of a watt of output but this can be very useful. It is very simple and uses a common integrated circuit. The output is low impedance and an 8-ohm speaker may be connected directly without the need of a matching transformer. A transformer is used to bring the impedance from the input to a high level which can easily drive the integrated circuit amplifier. Very few components are used outside of the integrated circuit, and all of them are non-critical and easily obtained. An amplifier of this nature requires strict adherance to proper wiring principles. All component and battery leads should be as short as possible to prevent oscillation of the circuit. High-gain amplifiers will easily break into oscillation and damage the integrated circuit when improper wiring techniques are used. Make certain all solder joints are made correctly, because a high resistance or cold solder joint can cause many circuit problems.

Control of the output volume is provided by a variable resistor, which is installed in parallel with pins 2 and 6 of the integrated circuit. A 9-volt transistor battery supplies power. This integrated circuit was designed for use with an external heat sink, but due to the low power consumed and dissipated by this device, it may be installed in dip IC socket without the need for external heat sinking.

To operate the circuit, first check to see that all wiring is correct. Make certain that the capacitor which connects the output of the integrated circuit to the 8-ohm speaker is wired while observing correct polarity. The positive terminal of the capacitor should be connected to the integrated circuit while the negative terminal connects directly to the speaker. Make sure the input transformer is also correctly connected. The blue and yellow wires are connected to the encoder. The red and white wires connect to the IC portion of the circuit.

The circuit current drain is sizable, and a single 9-volt battery will rapidly deteriorate. For portable operation, it may be desirable to use several of these batteries wired in a parallel configuration. This still provides a 9-volt output but the batteries will last longer by a multiple of the number of the batteries used. A

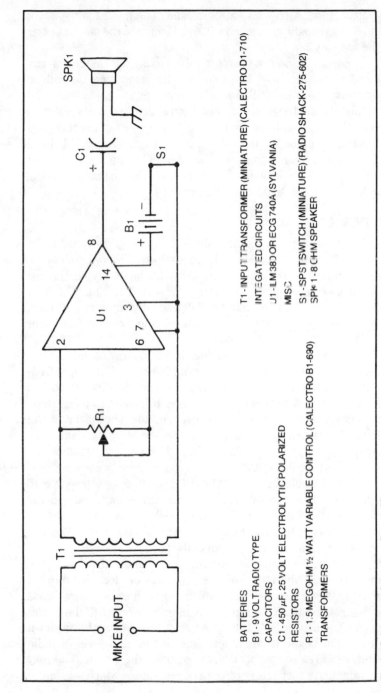

BATTERIES
B1 - 9 VOLT RADIO TYPE
CAPACITORS
C1 - 450 µF, 25 VOLT ELECTROLYTIC POLARIZED
RESISTORS
R1 - 1.5 MEGOHM ½ WATT VARIABLE CONTROL (CALECTRO B1-690)
TRANSFORMERS

T1 - INPUT TRANSFORMER (MINIATURE) (CALECTRO D1-710)
INTEGRATED CIRCUITS
U1 - LM 380 OR ECG 740A (SYLVANIA)
MISC
S1 - SPST SWITCH (MINIATURE) (RADIO SHACK-275-602)
SPK 1 - 8 OHM SPEAKER

Fig. 8-12. Integrated circuit audio amplifier.

235

12-volt lantern battery may also be used. The higher voltage will do no damage to the circuit and will even increase the output power to the speaker.

Connect the power source to the circuit and listen for any signs of oscillation. If a steady tone is heard in the speaker, with no input, this indicates oscillation and the power should be removed immediately and the wiring checked and shortened if necessary to prevent integrated circuit damage. If the first part of the test goes correctly, simply adjust the volume control for an acceptable level while receiving input from the encoder. The input transformer must have a high impedance secondary. The primary winding should match the output from the encoder.

ANOTHER IC AMPLIFIER

When more audio output is desired, a higher-powered audio amplifier must be built which will boost the audio volume level of the input tone from the encoder. Figure 8-13 shows a typical circuit which uses the Motorola MFC9010 integrated circuit. This is a more complicated circuit, but it will deliver slightly over 2 watts of audio output power, many times that of the previous amplifier project.

The input from the tone encoder is fed to pin 7 of the IC. The input is low impedance, on the order of 120 ohms. The output from the two-transistor tone encoder may be applied directly to this circuit in most instances without having to resort to a matching transformer. If other encoders are used, the secondary of the matching transformer should have an impedance of 100 ohms with the primary wound to exhibit an impedance which matches the output of the tone generator.

Wiring should be kept as short and direct as possible to prevent oscillation in this high-gain circuit. Again, due to the low-impedance output of the integrated circuit, no matching transformer is needed for the speaker which is coupled through an electrolytic capacitor to the output of the IC.

This circuit can also be constructed on vector board using the same mounting where the soldered contacts lie does not come in contact with any metal surfaces which might short out power leads and destroy the IC. No volume control is included with the circuit although one may be added at the output of the encoder to cut-down on the input of the IC amplifier. Due to the high level of audio output, a twelve volt supply is used to power the circuit. This may be obtained from batteries or from a power supply which acts on the

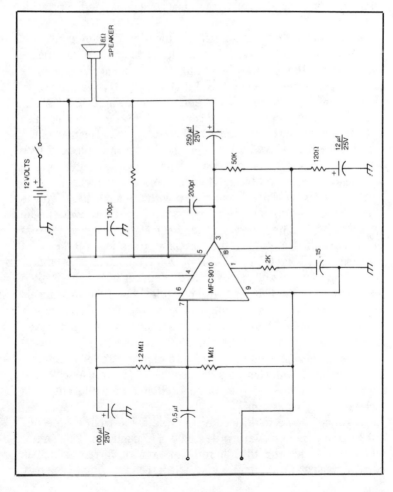

Fig. 8-13. A 2-watt output audio amplifier using the Motorola - MFC9010 integrated circuit.

237

transformed and rectified line current. This should be well regulated as should all supplies which power integrated circuit devices.

TONE DECODERS

We have already discussed some of the aspects of decoding the received tone signals through use of LC filters which conduct only tones within the range of their pass frequencies. This electrical signal is then passed on to the actual switching circuit which is comprised of a transistor, SCR, triac or some other type of solid state device which, in turn, may trigger a latching relay. The use of inductors and filters is quite simple, but determining their correct values may involve the use of complicated electronic formulas.

Another way of accomplishing the decoding of generated signals is through the use of a phase-locked loop. A phase-locked loop, up until 10 years ago, would have taken a massive amount of discrete components to build. Solid-state circuitry and integrated circuits have taken most of the size and complexities out of phase-locked loop design, and circuits can now be built which use only one integrated circuit and a minimum of discrete components.

Figure 8-14 shows a block diagram of a phase-locked loop which consists of a phase comparator, low-pass filter, and a voltage-controlled oscillator. The latter block of the loop provides a reference signal which is compared with the input tone. This is why the phase comparator bears its name. Its job is to *compare* the two signals. The imbalance of these signals at the input is what determines the output from the comparator section of the PLL. If the input signal and the one from the voltage-controlled oscillator are locked together and are exactly 90 degrees out of phase, then no output will be delivered to the amplifier, but if the two signals are not locked together, an imbalance occurs, and the comparator passes an electrical signal. The circuit, then, works on a prearranged *error*. The error is the difference between the input signal from the tone encoder and that of the voltage-controlled oscillator. The error signal will change the phase of and the frequency of the signal from the voltage-controlled oscillator until the output is locked with the input signal.

A very simple tone decoder can be made from a type 567 integrated circuit for a very wide range of frequencies. This circuit will safely handle up to 100 milliamperes of current which is certainly adequate to trip most small relays. The frequency response can be tailored by the selection of external, discrete components to fall anywhere between one hertz to well over 500 kHz. Figure 8-15 shows a block diagram of the internal circuitry of the IC which includes an output stage and a quadrature phase detector in addition to the other components of the former drawing.

The power output stage is triggered by the quadrature comparator and acts like a transistor switch. The transistor, however, is not a discrete component. It is contained on the single chip which makes up the integrated circuit.

Figure 8-16 shows a typical decoder circuit which is set up to trigger the internal transistor when it receives an input from the tone encoder of 1000 hertz. Components C1 and R1 determine the device response. Any other tones within its range may be selected by altering the values of these two components. The formulas for value determination:

$$R1 = \frac{1}{F \times C1}$$

where f is the pass frequency or triggering frequency. Resistance is given in ohms and capacitance in farads. R1 cannot drop below a value of 2000 ohms and should not exceed 18,000 ohms in this circuit. Capacitance values must be chosen which will allow the resistor value to stay within these confines. This is not difficult to do, and a lot of variations in capacitance and resistance may be used as long as the minimum and maximum resistance values are not exceeded. R1 will usually take the form of a variable resistor which may be set and re-set at will to vary the response of the triggering circuit.

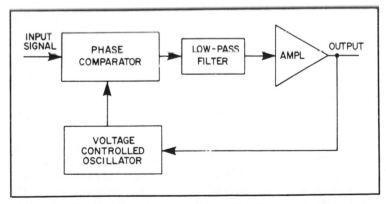

Fig. 8-14. Block diagram of phase-locked loop consisting of phase comparator, low-pass filter, and voltage-controlled oscillator.

239

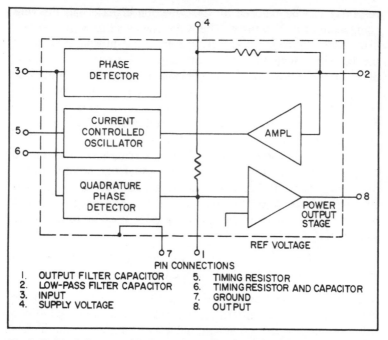

PIN CONNECTIONS

1.	OUTPUT FILTER CAPACITOR	5.	TIMING RESISTOR
2.	LOW-PASS FILTER CAPACITOR	6.	TIMING RESISTOR AND CAPACITOR
3.	INPUT	7.	GROUND
4.	SUPPLY VOLTAGE	8.	OUTPUT

Fig. 8-15. Block diagram of the internal circuitry of the 567 integrated circuit.

Looking again at the circuit, we see that a 0.4 microfarad capacitor is used for the value of C1. Taking this through the formula, we find that:

$$R1 = \frac{1}{1000 \times .0000004}$$

OR

$$R1 = \frac{1}{0.0004}$$

$$R1 = 2500 \text{ ohms}$$

A variable control which offers a resistance value of from zero to 5000 ohms will be ideal for this application, and will allow the frequency to be finely set, because the resistance value required falls at the center of its resistance range.

A smaller value control might be used by slightly modifying the circuit as shown in Fig. 8-17. Since the value of R1 cannot drop lower than 2000 ohms, a fixed resistor is placed at contact number 6. The control will be attached to the contact points marked "A" and "B." The formula now must take into consideration the fact that a resistance is already permanently wired and becomes active when the variable control is placed in the circuit. The formula would read:

$$R1 = \frac{1}{F \times C1} \quad \text{minus 2000 ohms}$$

For the example given, the variable control would equal 2500 ohms less 2000 ohms or 500 ohms. Now, a 1000-ohm control would be used. This latter circuit is electrically the same as the other, but the 2000 ohms minimum is already wired into the circuit. This prevents the possibility of the variable control being set for a too low resistance value.

The power supply for this device is a 9-volt battery which also powers the load. Small relays which will operate from 9 VDC at less than 80 milliamperes should be ideal loads for this circuit. The relay can then control the power to the work load.

The input signal would have to be received by a low output audio pickup, such as a microphone, and then passed through a

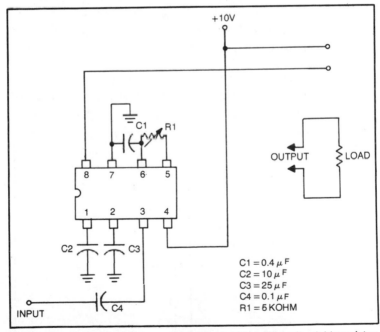

Fig. 8-16. Typical decoder circuit using integrated circuit and variable resistor for variable tone response.

241

small preamplifier. Audio pickups with higher outputs may be adequate to directly drive the PLL.

To make a multichannel version of this circuit, it will be necessary to duplicate for as many channels as are desired. The IC's are relatively low in price, and several may be purchased and wired in the same case. Construction is done on a small piece of perforated circuit board as was done with the other projects. Keep the wiring as short as possible and make sure of all connections and solder joints. The internal low-pass filter should prevent any oscillations from within the circuit and proper wiring technique will assure that this section does its job properly. The extra channels can use alternate values of R1 and C1 to provide multitriggering functions. Slight differences may exist between the frequency output of the tone encoder and the pass frequency of the decoder. Slight adjustments in the variable tone control should easily and quickly pull both systems into alignment. An oscilloscope can be useful when exact frequency setting is mandated.

SUMMARY

Tone-controlled systems are relatively easy to construct for a minimal cost in the home electronic workshop. Many of the coils needed can be purchased through surplus outlets. One inductor of great interest to the experimenter is the 88-millihenry toroid shown in Fig. 8-18. This consists of many turns of small diameter, enamel-coated wire wound on a core of ferrous iron. Using an iron instead of an air center magnifies the inductance values of the physical coil, and a large amount of inductance in a relatively small

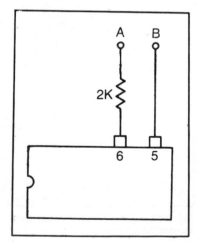

Fig. 8-17. Modification to former IC decoder circuit. Variable resistor is wired across points A and B.

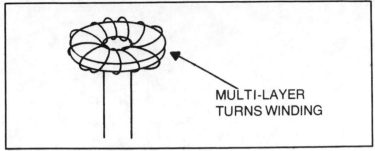

MULTI-LAYER
TURNS WINDING

Fig. 8-18. Toroidal coil wound on ferrous iron core.

space can be had. Many remote-control applications require the circuits to be as physically small as possible. These coils are ideal for such applications. They can be purchased very inexpensively and, when paralleled with capacitors of the proper values, they make excellent L-C filters.

The various electronic parts and integrated circuits can be purchased from your local Radio Shack or other such electronic hobbyist outlets. Construction is not highly critical nor are the component values in most cases. Slight deviations in values of 10 percent up to 20 percent down can usually be tolerated in tone circuits which offer a variable frequency adjustment, but, whenever possible, use the exact values and parts specified unless they are unobtainable in your area.

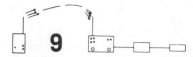

9

Light-Activated Systems

A popular method for controlling systems from a distance without resorting to hard wiring is through light-activation. This involves the focusing of a beam of light on a remote sensor which, in turn, switches control functions to the work load. Some applications involve the *removal* of a light source from the remote sensor. In the absence of the light beam, the remote control function is initiated.

Many of the components and circuits already described can be made to directly function as remote control triggers and decoders. The LASCRS, phototransistors, and photocells are basically all that are required to build a remote control, light-activated system. Some additional circuitry will be necessary, for certain applications, but all of this involves simple switching and control techniques which are easy compared to the mechanical functions which must be required to initiate and complete many of the types of work.

Figure 9-1 shows a circuit which uses a cadmium sulfide photocells (CdS) in conjunction with a variable resistor to trigger a small, latching relay. The cell exhibits a very high resistance on the order of half a million ohms when in complete darkness. This value drops to just a few ohms when the sensing window of the cell is struck by a light beam.

By referring to the diagram, it can be seen that this circuit is simplicity in itself. Only three major components are used, the cell, the variable resistor and the miniature relay. The power supply is not specified, but it can be anything from a 1.5 volt battery to a 12-volt, AC-derived circuit. The relay voltage will determine the value of the source voltage. The variable resistor is a commercial potentiometer with a value of about 50,000 ohms (50K ohms). This is used to set the activation level of the relay which should be a low-voltage device requiring less than 15 milliamperes of current to close. The CdS was purchased from Radio Shack for about $2. This component is noncritical, and most other types should work just as well.

Electrically, this circuit looks like a power source with a single resistor in series with the current flow. The CdS is really a variable resistor whose value is determined by the amount of light present at its optical window located on the top of the case. The series resistor which is variable by hand is set to latch the relay when the light source is trained on the optical window of the cell. After the relay has triggered, turn the light source off for a moment and then focus it on the cell again. The relay should now unlatch. If it does not, the variable resistor is set for too low a value and current is remaining on at all times. If a nonlatching relay is used with this circuit, testing is a little simpler. Shine the light on the cell, adjust the variable control for triggering. Now, remove the light from the optical window. The relay should immediately close. If it does not, re-adjust the control and try again until positive triggering is obtained. In some instances, it may be necessary to re-position the CdS or use a brighter source of light. Make certain the relay does not draw more than the recommended amount of circuit current. The CdS can be quickly destroyed by exceeding its maximum ratings. If a larger relay is needed, use separate latching relay and another voltage source which is triggered by the relay in this circuit. In this application, the latching relay should be replaced by the standard type which requires constant current to hold.

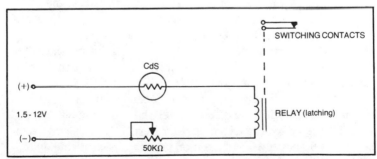

Fig. 9-1. Cadmium sulfide photocells circuit using adjustable resistors and low current relay.

This part of the circuit is simple. It is built on a very small piece of vector board and need not even be mounted in a protective case. The variable resistor can be mounted directly to this circuit board by drilling a hole at the board center, fitting through the shaft and the rads, and then by attaching the securing nut over the shaft.

If a case is used, some means must be provided for exposing the optical window of the CdS cell. This can be mounted to the outside of a metal case or cabinet as long as it is installed on an appropriate insulator which will not allow its case and leads to short out. If this should happen, no damage to the circuit will result, but the relay will receive a constant supply of current, and the control functions will be lost.

Another problem presents itself. The CdS cell must be wholly or partially shielded from ambient light which may increase in intensity as the sun climbs higher in the sky. Figure 9-2 shows a recommended method for CdS cell installation which involves enclosing it in a long, cardboard tube. It will be necessary to focus the light source down the length of the tube in order to actuate the circuit. If fairly constant, dim light levels are present in the vicinity of the CdS cell, this latter arrangement may not be necessary, as the variable resistor can be set higher to make up for the lower resistance of the cell due to the ambient light. This may be adequate to prevent false triggering but is not recommended for critical applications.

Figure 9-3 shows another circuit which uses a phototransistor to supply power to a relay. Actually the phototransistor drives a small npn transistor which performs the switching function. The circuit consists of Q1 which is a Radio Shack FPT 100 or any similar phototransistor type. When light strikes its optical window, it passes base drive to Q2 which is a 2N706A or its equivalent. A miniature relay which will operate from 9VDC is used as the actual controlled switch. This device may be mounted directly to the circuit board where the other components are located or may be brought out to a point closer to the work load through insulated leads. The power source is a nine volt battery although this circuit should work with any supply of between 6 and 12 volts.

Vector board construction is recommended, although this circuit is small enough to be completely built on a small terminal strip if the relay is located elsewhere. Alternately, a printed circuit board can be easily etched from one of the kits available at your hobby store. This latter design will provide the greatest amount of stability. It will also make the circuit fit into a smaller area.

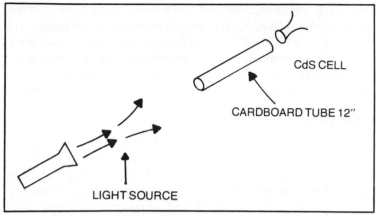

Fig. 9-2. Possible method for cadmium sulfide photocell installation using hollow, cardboard tube as a shield from ambient light.

The triggering of the phototransistor was not adjustable when this circuit was first made. It was not necessary as the ambient light level was not high. An adjustable circuit could be made by inserting a variable resistor in the lead between the collector of the phototransistor and its lead from the nine volt source. Experiment with the value to obtain the desired effect. You might start with a

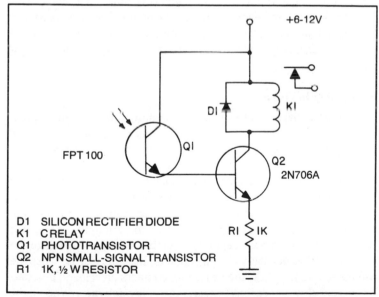

Fig. 9-3. Light-activated circuit using a phototransistor as a triggering device at the base of npn transistor, Q2.

247

2500-ohm potentiometer and work from there. For a fixed, noncontrollable circuit, the potentiometer can be temporarily inserted in the line and adjusted for the desired triggering level. After this is attained, remove the control from the circuit and measure its resistance with a multimeter. Replace the control with a fixed resistor close to the measured level of the potentiometer.

LIGHT SOURCES

Little has been said about the light source which will serve as the remote control command transmitter. Most often, this will be a multibattery flashlight or lantern which casts a bright, focused beam of narrow light, but almost any source can be used. The concentration here is on focus and narrowness of beam. Wide dispersion lights often will not provide enough concentrated light on the optic windows of the various optoelectronic devices.

Multichannel control is very simple by converting or, rather, multiplying to systems already discussed. The last project is turned into a multichannel control circuits simply by adding more relays, phototransistors, etc. and building matching circuits all of which operate from the same power supply. All circuits can be built on a single piece of vector board, but the optical windows must be adequately separated to prevent triggering more than one circuit with the same light beam. Again, a cone of metal or cardboard may be placed around each optical window to prevent accidental triggering from the control source or from changing ambient light conditions.

SOLAR CELL DEVICES

Solar cells are often used for simple types of automation control which reacts to the sunrise. Figure 9-4 shows a simple circuit which is often found in the day-night automatic switches many homeowners use on their outside lights when they are away. The circuit consists of four small solar cells which are wired in series and provide an output of a little over one and a half VDC when directly in the rays of the sun. They are mounted in a position where they will be directly exposed to the sunlight. When this happens, the cell circuit generates electrical power to engage the relay which has normally closed contacts. In other words, when the cell bank is in the dark, the switching contacts are closed because no current flow is present to make the relay activate. When light strikes the cells, the relay is engaged by the current flow and the switch opens. Normally, one leg of the 120-volt house current line is controlled by this switch. The line can supply operating current to lights or any other devices which do not draw more current than the relay contacts are designed to handle. In daylight the relay is activated, so the 120-volt circuit is broken, but at night, the relay is off, but the switching contacts are closed. Current flows through this contact from the house supply and powers the load.

LASCR CONTROL SWITCHES

Figure 9-5 shows a remote control switch using a LASCR circuit which is very simple and economical to construct. While large currents can be handled by a LASCR, this particular component is rated to handle about 1.5 amperes. This circuit can control a 150-watt incandescent light or any other circuit requiring 120 volts of DC operating current. Actually, the voltage will soar to nearly 150 VDC under light loads or no-load conditions. The LASCR used was a Radio Shack purchase and is designated as part number 276-1095 but virtually any LASCR will work in this circuit as long as it is rated to withstand the voltage and current levels. A 47K ohm resistor is attached between the cathode and the anode to filter the pulsating DC output from this circuit.

When light strikes the optical window of the device, current begins to flow through the load which can be almost any device requiring 120 volts DC. When the light is removed from the

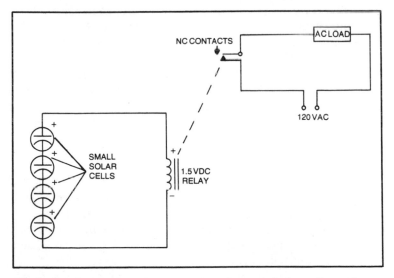

Fig. 9-4. Four small solar cells wired in series can control a low-voltage relay as ambient light changes.

249

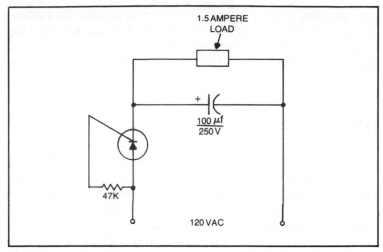

1.5 AMPERE
LOAD

$+$ |(
$\frac{100 \mu f}{250 V}$

47K

120 VAC

Fig. 9-5. Light activated control switch using a LASCR device and a few simple electronic components.

window, the LASCR stops conducting almost immediately and the circuit appears to be open when looking at it from the power source.

This circuit should be carefully wired and well insulated. It is imperative that it be mounted in a protective case to prevent accidental contact. House current is very dangerous. Do not take the chance of having an accident by leaving voltage points of this potential exposed to possible contact. Arrange the LASCR so that it is easily reached by the narrow beam of light which is used as the remote control sender. For AC control of motors and the like, a light-activated triac may be substituted for the LASCR or two LASCRs may be wired in triac fashion as was described in an earlier chapter. Make certain that the components are rated to withstand the calculated surge current ratings of the motor if this type of device is to receive power from the circuit.

OPTICAL CONSIDERATIONS

An effective system for light-activated remote control will almost always involve the addition of optics to transmit or to receive the controlling light beam. To assure positive triggering, this is often necessary, because changing ambient light conditions can cause a very positive triggering circuit to be quickly relegated to partial and unstable operation.

For the reasons given, many experimenters prefer to add filters to their light-activated systems. These filters determine the type or types of light which will cause the various electronic circuits to trigger. Some of the devices on the commercial electronics and hobby market are already designed to respond best to different light frequencies. These are often more expensive but also provide more positive triggering action. Perhaps, your system might respond best to infra red light. This will depend upon the exact conditions and general environment of the control area. If the light sensor at the work load will be exposed to the direct rays of the sun, it will be very difficult to build any kind of system which will offer positive triggering because of the intensity and variation of light levels which will be present at the sensor. In a case such as this, light-activation might best be discarded for one of the many other triggering methods which are not plagued by ambient light problems.

On the other hand, if your sensor and work unit are to be placed inside the home and at a spot near the remote control point, light activation may be one of the easiest ways to go especially if you desire to control only the "on" "off" functioning of a circuit. For instance, light-activated switches work very nicely with indoor television receivers. These appliances are normally viewed in subdued lighting, and the remote operator is certainly close by. Through means of a small flashlight at the remote control, the television may be switched on or off with any of the light-activated controls discussed here. Other functions such as channel change and volume control could also be included by installing other light channels and adding a stepping switch could increase the volume while the motor would be ganged to the channel selector to effect these controls from a remote point of viewing.

It can be seen that effectively getting the light to the remote sensor is the main problem with light-activated control systems. Some rather elaborate optical systems have been built which will allow the light from the control point to be finely focused on the optical window of the solid-state sensor. These have worked fairly well, but often this increased the sensitivity to ambient light fluctuations, because this light was also intensified by the optical system.

Sometimes it is desirable to completely enclose the light sensing circuit and run light beams through tubes, just as electrical current flows through copper conductors. The light conductors are popularly called fiberoptics. They can be purchased from hobby stores and catalogs for quite reasonable prices for small lengths of the material. A fiberoptic is a small, slender tube which is flexible and can be bent and twisted for routing through electrical circuits, around corners, and anywhere necessary to establish a light circuit. When light is applied at the input of the fiberoptic, it is conducted through the length of the tube to the output. The light output is correct and corresponds to the light at the input no matter how many twists and turns occurred in the fiberoptic circuit path. These light cables are now replacing hard-conductor phone cables for transmitting voices through light modulation.

Through fiberoptic circuits, more stable and dependable control using light activation can be had. The distance between the light source and the light sensor is gapped by the fiberoptic conductors. This may seem odd, because we have reverted in some ways to a hard-wired or direct control system instead of being completely isolated from the work load or control source. This is true, but this system has many advantages especially in commercial high voltage applications. It can also be used at home for triggering slave flash units from the flash of your camera-mounted unit.

While this system does provide for physical connections between the remote control point and the work load (actually, the connection is not direct as there is a small gap between the output of the fiberoptic and the LASCR or other control device). It also offers the most positive type of triggering. The LASCR, CdS, or phototransistor can be completely housed inside a light-proof cabinet. The only source of light for these devices will be from the output of the fiberoptic system. Now multichannel remote control becomes very practical, because many many optical strands can be tied together in a cable arrangement and conveniently run to the sensor control point. These optical pipes are so small and slender, that many of them can be run in parallel without getting bulky.

Figure 9-6 shows a multichannel light-activated arrangement which uses LASCRs as triggering devices. While only five channels are shown, many more could be added by increasing the number of fiberoptic strands and adding additional LASCR circuits. The LASCR in each circuit is closely coupled to the output end of the fiberoptic which acts as its control line. The flexible optic tube can be adjusted for best activation of the LASCR.

This system works well for high voltage applications where no direct connection between the remote operator and the high-voltage system should exist. Fiberoptics are conductors of light rays but not of electrical current as we think of it. The fact that the optics do not physically connect to the voltage and current control portions of the circuit (this is handled by the LASCRs) is another safety advantage.

Figure 9-7 shows a commercial circuit supplied by *Motorola Semiconductor Products Inc.* which shows a light-activated, high-voltage series switch capable of switching a power source of 60,000 peak volts. The circuit is composed of a string of SCRs in series which is fired by phototransistors driven by a xenon flash tube. A bundle of fiberoptics is used to conduct the light from the flash tube to the optical windows of the phototransistors. The fiberoptic bundle is a commercial unit which splits its input into 10 equal output channels as shown in the schematic drawing. Several additional components have been added to each of the ten switching circuits over what was included in earlier circuits of this type. The 1.5M ohm resistor in series with a 51K ohm resistor form the voltage equalization network for the series connected SCRs and also provide a voltage divider network for each switching unit.

When the flash tube is activated, the light enters the input of the fiberoptic bundle and is split into ten outputs, each of which is channeled to the sensor of the phototransistor. The output at each transistor is balanced, so each will receive a burst of light which is approximately equal in amplitude to the bursts applied at all other phototransistors in the circuit. As the light bursts strike, each phototransistor conducts and discharges a 0.1-μf capacitor into the gate of its corresponding SCR through a 510-ohm resistor. Since this happens ten times at once in this circuit and at all SCR

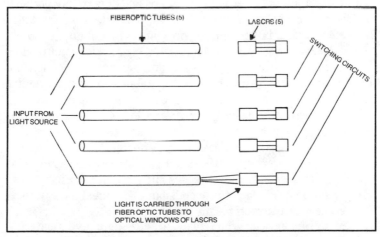

Fig. 9-6. Multichannel LASCR system with separate fiberoptic conductors used as control elements.

switches, the entire string of SCR's turn on simultaneously. Notice that no direct connections between the input light source and the input to the phototransistors has been made. An extremely high voltage source has been controlled from a remote location by conductors of light energy instead of standard electrical current.

SUMMARY

Light-activated remote control systems offer several advantages for noncritical uses around the home, however, if highly accurate and dependable systems are to be designed for critical control functions, the price can quickly soar because of the physical conditions which must be observed. When a large system uses fiberoptics as light conductors, the price of that system can multiply by staggering amounts. Fiberoptics are relatively inexpensive when purchased in short bundles, but these are simply manufacturer's cuttings and leftovers which are sold to the hobby suppliers. If you need a long fiberoptic cable, the price in small quantities can be quite prohibitive.

Light-activated remote control usually operates most efficiently in simple systems contained within a room of the home. Most of the systems that can built at home will not lend themselves to long-range, especially outdoors in the daylight hours.

Their convenience around the home, however, is a matter of record, and the builder need only wire a few simple SCR or LASCR circuits as sensing units to control a relay or directly power a DC device. The sending unit used at the control point is nothing more than a flashlight or lantern in most cases and is certainly portable and readily available at all times. The adjustment of the light-activated system is very simple, in that no tones must be finely tuned to exact frequency, no complicated electronic circuitry need by matched between control point and sensing point, and the overall cost is usually lower than comparable short-range systems which use more complex electronics to form their circuits.

Light-activation leaves room for a lot of experimentation, especially with optical systems, in achieving longer ranges in daylight conditions. There are several hobby and experimenter outlets which advertise in the national electronic magazines that specialize in a wide variety of lenses and lens accessories. These products are usually factory seconds but make ideal lens systems for remote control experimentation.

Many other circuits can be used to sense light levels and they, in turn, can control almost all of the working systems discussed in this text. There are so many different types of optical and

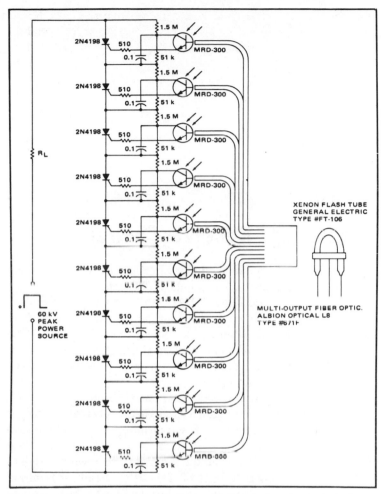

Fig. 9-7. Commercial circuit using light-activated devices for control of very high-voltage source.

optoelectronic devices on todays market that it pays to keep up with what is being offered. These are all basically the same types of devices, but with modifications, filtering systems, etc., so instead of attempting to "rig" a circuit from available parts, it might be a good idea to check around and see if what you need has already been invented and marked. This is not meant to discourage expermentation and homebuilding which is highly recommended, but when a certain component is needed for a remote control system, this can save many hours of work.

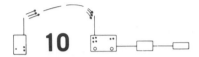

10

Transmitter-Receiver Control

One of the most efficient and practical means of establishing remote control from distant points is through means of a high frequency transmitted signal. Many commercial transmitters and receivers are available on the communications and hobby markets which can be directly adapted for remote control purposes. The distance of transmission can vary from an effective range of a few hundred feet, to several miles, to thousands of miles, depending on the overall design of the system.

Figure 10-1 shows a block diagram of a basic system which uses the activation of the squelch circuit in the receiver as a means of controlling work functions. The transmitter is set to broadcast on the receiver frequency. A squelch circuit within the receiver removes all audio output from the sampling circuit which drives a solid-state switching relay control. This latter circuit can be any of the ones previously discussed, although it will not be necessary to design any type of tone filter.

When the transmitter is keyed, the receiver squelch activates the audio output of the receiver which amplifies the internal noise of the receiver when a transmitted carrier is present without any modulation. If the receiver audio gain control is set to a high output point, the generated noise signal should be adequate to drive the simple switching circuit which might consist of an SCR coupled to the receiver output through a small transformer.

The switching circuit can drive a latching relay which will switch operating current directly to the load. This circuit has been discussed in detail in previous chapters and will not be dealt with

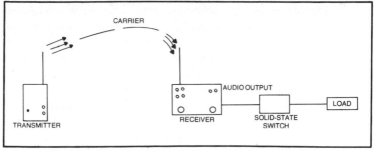

Fig. 10-1. Block diagram of basic, transmitter-receiver control system.

here to avoid repetiton. This simple circuit can be made to work with almost any type of transmitter-receiver combination. The effective range will depend on the power output of the remote control transmitter, the sensitivity of the receiver, the frequency, and the antennas used for transmission and reception.

Figure 10-2 shows two children's walkie-talkies which were set up for remote control applications. The pair costs less than thirty dollars and will provide remote control over a distance of about 100 feet with the small, whip antennas built into each unit. Both have the ability to send and receive, but one is used solely as a transmitter for these applications while the other is used for its receiving ability.

Fig. 10-2. Children's walkie-talkie can be purchased cheaply and used for remote control over short distances.

Most of these transceivers are on the Citizen's Band channels, but many of the newer models operate on a fairly new frequency allocation around the 48-megahertz region of the radio spectrum. These latter units are to be preferred for remote control applications, because they are far less crowded with users. The remote system will function anytime the squelch of the receiver is broken by a transmitted signal on its operating frequency. Band crowding means that the squelch would be broken quite often, and undesired triggering of the switching circuit would be the result. This can be overcome in some instances by internally lowering the sensitivity of the receiver. This can also be done by simply shortening the receive antenna of a walkie-talkie by telescoping the antenna completely into its case. This lowered sensitivity will mean that only transmitting stations very near to the receiver will be able to break the squelch control.

Sometimes, when very good receivers are used, the internal noise is insufficient to drive the relay switching circuit. Here, it will be necessary to speak into the microphone of the transmitter to provide an audio output at the receiver for the switching circuit to act upon. A switching circuit which can be adjusted for sensitivity is shown in Fig. 10-3. It is designed to operate from the receiver output which is sampled through the external speaker jack located on most units and is engaged by means of a miniature male plug. The output signal is amplified by a simple, solid-state amplifier circuit which feeds its output to a transistor switch. When the transistor receives a base driving signal, it conducts and the relay is engaged by the power supply which can be a battery in the six to nine volt range. This, in turn, can provide electrical power to a larger latching relay which will be engaged with one burst of power and disengaged with another. This means that the transmitter and receiver need be activated for only a second or so to start operation with the same time of use needed for stopping the power to the work load.

The components for this circuit are relatively noncritical and can be purchased at your local hobby outlet. Construction is done on a small piece of circuit board. The finished board is mounted in a small, aluminum box which is grounded to the case of the receiver if possible. Larger receivers may even enable the builder to mount the finished circuit board within the terminal strip on the back of the unit. A switching function can be added to a receiver in this manner without increasing the physical size or appearance of the device.

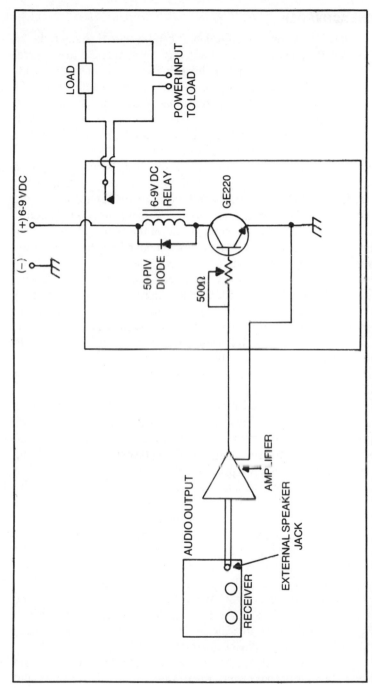

Fig. 10-3. Switching circuit designed to be placed at the receiver's audio output for remote control.

259

TONE ACTIVATION

In using the previous sytems, it can be seen that any time the receiver frequency is keyed by *any* transmitter, the control circuitry will be activated if modulation is present on the transmitted signal. With the active use of most frequencies today, this presents problems with unintentional activation. In the more critical control areas, it is necessary to design a system which transmits coded command signals in order to assure controlled activation through positive keying actions.

The same, basic systems as before can be used for these purposes, but the receiving system must be setup to trigger the switching and control circuits only upon receiving a preset audio signal or signals which are sent by the transmitting unit located at the remote control keying point. Most of the time, this triggering signal is an audio tone which is superimposed on the transmitted carrier by amplitude modulation. For a very basic transmitting system, all that is necessary is to feed the audio output from a signal generator into the microphone input of the transmitter. A more direct approach can be had by feeding the output of the generator directly into the audio input of the transmitter modulator. This bypasses the need for a microphone and prevents extraneous audio signals at the microphone from being transmitted to the receiver, possibly causing false triggering or no triggering at all.

Figure 10-4 shows a block diagram of this system which uses many of the circuits discussed in previous chapters. Following the operational path from left to right, the tone generator output is fed directly into the microphone jack of the transmitter. A matching transformer is used to provide distortion-free coupling. The primary of this transformer matches the output impedance of the generator's output, while the secondary is the same impedance of the microphone. The audio tone is superimposed on the carrier of the transmitted signal which is intercepted by the receiver antenna. The receiver is tuned to the transmitter frequency and detects the tone which it reproduces at its audio output. This output is connected to a filter whose pass frequency is the same as the tone from the generator. When the signal is passed on to the switching circuit, it is in the form of an electrical current which triggers the firing of an SCR or the conduction of a transistor, whichever solid-state device is used for switching.

This is an excellent system, because it will properly respond to only the tone or tones which the filter is designed to pass. Should

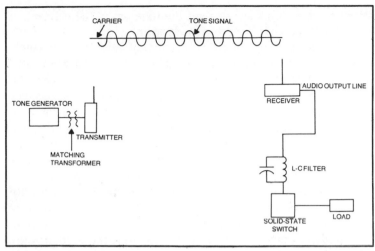

Fig. 10-4. Tone-activated system through modulated radio frequency transmission.

another transmitting station send on the receiver frequency, the circuit will not be triggered unless a tone of the same frequency as the filter pass-band is transmitted. This circuit should eliminate over 90 percent of the false triggering which can result with the carrier-activated systems that preceded this one.

The filter can be an LC circuit or can be designed around a phase-locked loop. Remember, the transmitter signal merely carries the tone from the audio generator to the switching circuit. After the carrier has been demodulated by the receiver, the resultant output is back in the audio frequency spectrum again. From this point on, the rest of the system can be handled in exactly the same manner as the audio tone systems of an earlier chapter. This is really a tone system, but the transmitter is used to make the audio tones travel a much further distance. The transmitter multiplies the audio frequency many times. If the tone from the generator is 1000 hertz, after passing through the transmitter it will be sent as a frequency of several million hertz, but when the receiver detects the transmitted signal, it separates the 1000-hertz tone at the receiver output.

It can be seen that an audio tone is the main controlling signal as it is the only signal passed on to the switching circuit. The 1000-hertz tone at the generator is passed directly on as a 1000-hertz tone at the input to the triggering circuit. The transmitter provided a vehicle for the tone to travel on while the receiver acted as a source to get the tone out of the carrier and allow it to travel on in its normal mode again.

Multichannel operation is easily obtained from a single transmitter and receiver by providing a tone generator with several, different frequency outputs. The receiver needs no modifications nor does the transmitter. The output of the receiver would then be coupled to several filter circuits, one for each channel of operation. If a four-channel system was desired, the tone generator would have to be capable of producing four, separate audio tones. Four filters would be used at the receiver audio output along with four switching circuits. Figure 10-5 shows a block diagram of such a system which provides four, separate control functions. There is really no limit to the member of channels which can be built into this system. Tones should be separated as far in frequency as possible to make the building of the filter assemblies less complicated. Filters are not perfect, and one which was designed with a pass frequency may allow frequencies 200 or more hertz on either side (up and down) of the center frequency to be passed on to the switching circuits with enough amplitude to activate the solid-state switches.

THE FEDERAL COMMUNICATIONS COMMISSION AND LICENSING

When remote control systems use transmitters, the Federal Communications Commission must be a consideration. This is the government agency which controls all radio communications in the United States. Certain types of remote control transmitters will require an FCC license before use. This will depend upon the type of transmitter, the operating frequency or frequencies, the power output and where and how the transmitter is to be used. The transmitters discussed so far in this chapter have all been the low-power walkie-talkies which do not require an FCC license as long as the units have been purchased from a company whose design has been approved and licensed by the FCC. These transmitters operate with a power input of 100 milliwatts (a milliwatt is one one thousandth of a watt) or less and whose antennas do not exceed the maximum allowed 5-foot length. These transmitters may be used directly for remote control work on any of the frequencies they are designed for. They are licensed under Part 15 of the FCC rules and regulations regarding Citizens Band operation.

Another class of operation has been designated by the FCC for use in remote control activities only. This is called Class C and differs from Class D, which is what most of the CB transceivers of today operate under. The many commercial remote systems made today for control of modern airplanes, cars, etc., operate under the Class C rules and regulations. In most cases, it will be necessary to obtain a license from the Federal Communications Commission to operate legally with these transmitters. This license is very easy to get and usually involves the filling out of a license form enclosed with all of these commercial models. If you build a remote control transmitter under any part of the FCC rules and regulations governing Citizens Band activity, it will be necessary to have your constructed circuit certified by a holder of an FCC Second Class Radiotelephone license. This makes certain that the transmitter output is within acceptable limits and that the frequency of operation does not fall outside of the allocated band of frequencies for Citizen's Band Control.

Remote control operators who have an Amateur Radio License may operate radio control transmitters on some six meter bands which start at 50 MHz. Operation here is detailed by the FCC and is far too involved to go deeply into within the confines of this text. A complete list of the various rules and regulations governing remote radio control can be had by writing the FCC Field Engineer in a city near you or by writing Federal Communications Commission, Washington, DC.

VARIABLE TRANSMIT FREQUENCY CONTROL

If you are licensed to operate a transmitter whose frequency can be directly controlled through means of a stable variable frequency oscillator, an effective means of tone control can be had without resorting to the generation of audio tones at the input to the transmitter. Amateur radio operators should be able to take advantage of this system.

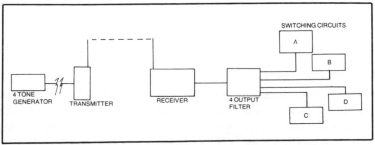

Fig. 10-5. Multichannel operation of modulated transmitter carrier system.

When a receiver detects a transmitter carrier on its own frequency, it demodulates the signal and converts it back into audio at the receiver output. The human voice is a complex conglomeration of audio tones which usually lie somewhere between 300 and 3000 hertz, depending upon the voice characteristics of the person talking. When no speech or modulation is present on the transmitted carrier, there is no signal to de-modulate, so only silence is heard from the receiver along with some of the internal noise and static which may be generated within the receiver circuitry. Amateur code transmissions do not contain any modulating waveform. The carrier is used in relationship to a bfo (beat frequency oscillator) to produce a tone at the audio output of the receiver. The bfo frequency is adjustable and is set near the frequency of the transmitted signal but not exactly on this frequency. When two signals are combined, the resultant output in the receiver is the sum and the difference of both. If a constant carrier was transmitted on a frequency of 27,215 MHz this would be the equivalent of a frequency of 27,215,000 hertz. Now, if the bfo is set to generate a frequency of 27,214 MHz, its equivalent of 27,214,000 hertz would be combined with the carrier frequency. The difference between these two signals would be 1000 hertz which is in the audio frequency range. A 1000 hertz tone would be heard at the output of the receiver for as long as the carrier was transmitted. This is the difference of the two signals. The sum of the two is also produced but lies far outside of the normal receiving range (in the vicinity of 54,429 MHz) and is not passed on to other circuits.

If a tone filter is placed at the audio output of the receiver and set to pass a 1000 hertz audio tone on to the switching circuitry, tone control has been obtained without having to generate a like tone at the transmitter input. This makes for a less complicated procedure, and due to this, the system will be more dependable and efficient.

Now, suppose you wish this system to be of multichannel design. This is very simple if the transmitter output frequency can be raised or lowered accordingly. Figure 10-6 shows a block diagram of the circuit which changes only, as to, the number of triggering circuits used at the receiver output. This example shows a two channel system, but it could be set up for 200 or more channels if this amount of circuit complexity was necessary for a very elaborate control system. Two audio tones have been chosen, one for controlling each solid-state switch. The first tone is 1000 hertz, and the second is 2000 hertz.

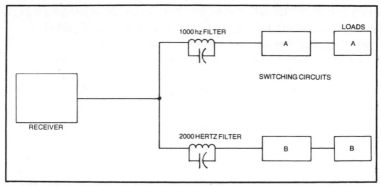

Fig. 10-6. Multichannel variable frequency control system.

For triggering the first tone, the frequency of the transmitter can be 27,215 MHz and the bfo frequency, 27,214 as was in the previous example. We know that when the two signals combine, they will produce a 1000 hertz tone at the receiver audio output. Now, to produce the second tone of 2000 hertz, the transmit frequency is changed to 27,216 MHz. The bfo frequency is left unchanged. The difference between the transmitted frequency and that of the bfo is now 2000 hertz, and a tone of this frequency will be present at the receiver output. The second filter passes this signal because it has been designed to exhibit a pass frequency of 2000 hertz, or 2 kilohertz.

This frequency change can be obtained with a Part 15 transmitter which is always crystal controlled if it has a multi-channel switch allowing it to transmit on the frequency of a second crystal installed in its circuitry. If the first crystal is chosen to produce a transmitted frequency output of 27,215, the second crystal would be chosen for 27,216 for the particular control system under discussion. If your transceiver had five-channel selection, then you could have five-channel remote control capability as well. Through all of the frequency changing, the bfo remained tuned to the same frequency, because the transmitted carrier was being raised and lowered in relationship to the bfo frequency.

Bfo circuits are found in communications receivers. Many of the cheaper receivers do not offer this feature as they are designed to receive AM transmissions only and cannot copy single-sideband and code signals. If your receiver does not have a bfo, one can be easily made which will provide a beat frequency by placing it close to the receiver antenna input. No direct connection is usually necessary. Most beat-frequency oscillators are of the variable type, but these are often very unstable and will allow the tone at the audio output to "travel" quite a bit. If a narrow filtering system is used, the received tone may fall outside of the passband of your filter. For this reason, the bfo shown in Fig. 10-7 is a crystal-controlled type which locks the device on frequency. This oscillator should work well with all crystals designed to produce harmonic outputs within the 3 to 30 MHz portion of the frequency spectrum. It consists of a single transistor, two capacitors, two resistors, a 215-millihenry rf choke and a crystal socket. A third capacitor is used at the sprocket to bring the oscillator onto exact frequency. This capacitor should be of the "trimmer" variety and should be mechanically secured in the circuit so that no physical movement can occur. Once the crystal is located in its socket, power is supplied from a 9-volt battery, although 6 to 12 volts should also work well. This power can be obtained directly from the receiver in many instances and should be regulated to avoid frequency drift. The receiver is tuned to the transmitter frequency which should be less than 2500 hertz from the intended frequency of the oscillator. Now, adjust the trimming capacitor until a tone is heard at the receiver output. If nothing is heard, place the oscillator rf choke closer to the receiver circuit. It will probably be necessary to compare the received tone to that of a tone standard set to produce a tone of the same frequency as the *desired* receiver output tone. As the two signals approach a match, you will hear a wavering or oscillation with your ear. When this stops and the two tones seem to be one, the circuit is properly aligned. By varying the transmitter frequency, a wide range of audio tones can be obtained at the receiver output to drive many filter/switching channels. Most Citizens Band crystals for operation on the frequencies discussed will be third-overtone types which mean that they will be designed to oscillate at one-third of the oscillator output. A crystal to oscillate at 27.215 MHz would be labeled as a 6.803 MHz or 68,030 kHz component. To arrive at this figure, divide the desired output frequency by 4. Choose a crystal cut for this frequency. The resultant output will not be an *exact* multiple of the crystal base frequency. This is the reason for the crystal trimmer control which can effect small changes in output frequency.

RANGE

The transmitter range which will determine the maximum distance the control point can be from the receiver and work load will vary. The maximum distance will depend upon transmitter output, antenna length and height above terrain. For most control

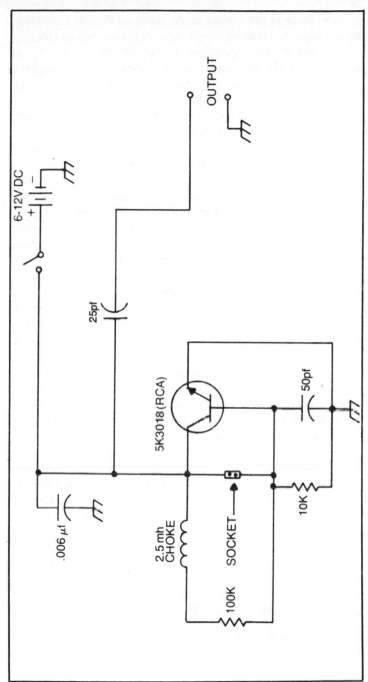

Fig. 10-7. Beat-frequency oscillator circuit.

operators, there is little that can be done, legally, to a commercially manufactured transmitter to improve range. Antenna lengths are often fixed by FCC regulations, so the obvious place to start is at the receiver and receive antenna. There are no restrictions here so the antenna can be of any length allowed by local regulations. Its height may be limited by the laws of your town or city charter.

The receive antenna should be of the same characteristic design as that of the transmitting antenna which, in most cases, will be a vertical radiator. This means that the antenna element lies in a vertical plane as compared to the plane of the earth. The receiving antenna should also lie in the vertical plane to get the most receive efficiency from the transmitter/receiver combination.

If control frequencies within the CB spectrum are to be used, any of the roof mounted or mobile antennas used for transceive operation may be attached directly to the input of your communications receiver. This will provide maximum receiving sensitivity. Alternately, a vertical antenna can be made. Figure 10-8 shows the basic design. The antenna is made from a length of small-diameter, lightweight aluminum tubing. Coaxial cable is used to feed the system. The center conductor of the cable is soldered to the bottom of the element. This is best accomplished by drilling a small diameter hole through the base of the tubing and attaching a bolt and nut at this point. The solder connection can be made directly to the protruding end of the bolt or the cable conductor can be wrapped around the bolt and held securely by the nut. Treat this connection with a weatherproofing compound if the antenna is to be mounted outside.

The braid portion of the coaxial cable is attached to a good ground system. This can be an earth ground made by driving a metal rod about 6 feet into the earth. Alternately, a radial system can be made from lengths of copper or aluminum conductor laid about the base in spoke-like pattern. This is a good arrangement for roof-mounting, because the ground system radials can be secured to the roof as shown in Fig. 10-9. If the antenna is to be installed over a tin roof, a connection made directly to the metal sheeting can serve as an excellent ground system. It is necessary, however, that the vertical radiator be insulated from the roof at all times.

The length for the vertical element is determined by the formula:

$$L = \frac{234}{f}$$

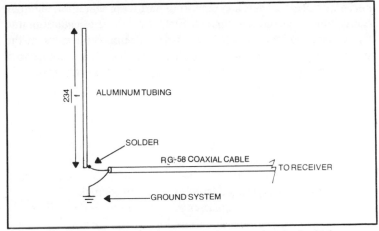

Fig. 10-8. Vertical antenna design using a quarter-wavelength element.

Here, antenna length (L) is equal to the figure 234 divided by the receive frequency in megahertz. The length of the vertical radiator for a receive frequency of 27.215 MHz would fit into the formula thusly:

$$L = 234/27.215, \text{ or } L = 8.958 \text{ feet}$$

The antenna length would be 8.958 feet or 8 feet 11½ inches, approximately. Any other frequencies of operation may be inserted for the correct antenna lengths using this formula. If a conductor ground system is to be used instead of an earth ground, three radials should be sufficient if placed equally around the base of the vertical element where the coaxial connection was made. The length of these radials will be the same as the length of the vertical element. Figure 10-10 shows how the ground system should be arranged. This drawing is an overhead view with the vertical element shown at the radial system center.

If this type of system is too complex for the mounting space, an alternate antenna can be used which should also provide good performance. This is a long wire antenna and should be mounted with its far end as high above ground as possible. Shown in Fig. 10-11, one of the antennas is connected directly to the receiver terminal while the other extends away from the receiver, out of a window, if possible, and up on the roof or to a distant tree. The element should be insulated from all metal. Insulated copper wire can be used for the element. The 300-ohm stand-off insulators which are often used for stringing twin-lead television cable can serve as excellent antenna mounting hardware when the side of a house must be used for support. Simply screw the insulators into the wood framing, and push the wire through the center of the insulating portion of this device. The length of the antenna wire should be twice that obtained by using the previous formula or:

$$L = \frac{468}{f}$$

Again, the length is stated in feet and the frequency must be introduced into this formula in megahertz. The only factor that has changed is the figure which has been doubled to arrive at a length factor of twice that obtained with the previous formula.

As far as receiver sensitivity goes, the better receivers are almost always more sensitive than the cheaper makes. Fortunately, most remote control applications do not involve great distances, and most receivers will be found to provide adequate sensitivity to detect the transmitted signals from the remote control point. Completely re-aligning an older receiver will probably result in a sensitivity increase, especially if alignment procedures have not been done recently. If a tube-type receiver is used, tube replacement will bring up the sensitivity if they have

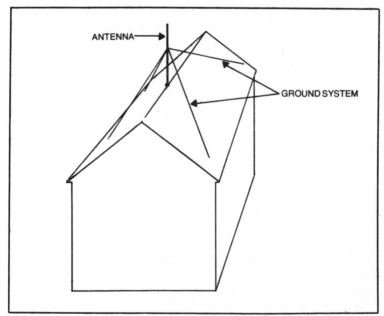

Fig. 10-9. Roof mounting of vertical antenna and radials.

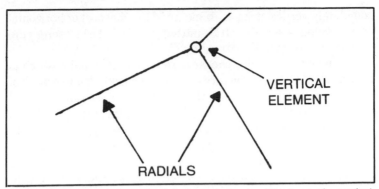

Fig. 10-10. Overhead view of radial placement around the base of a vertical antenna.

been used for many years without being tested. The rf amplifier tube is the main one to be considered when the receive sensitivity suddenly drops. If replacement does not correct the situation check all of the other tubes, looking for those with weak outputs as indicated on a good tube tester.

If your receiver seems to be detecting the transmitted signal, but the switching circuit is not triggering properly, frequency response of the audio circuit may be a problem, assuming that the filter and switching circuit are correct and set properly. Many communications receivers offer the best audio response between 300 and 3000 hertz. If you should use a control tone at the output above or below this range, the audio output or audio response to these tones may be suppressed. The result is an inadequate or intermittant signal output to the switching filters. This can be corrected by choosing audio control tones which lie within the frequency response range of your receiver. Some internal adjustment or modifications to the receiver circuitry may be attempted, depending upon the circuit design. The owner's manual and schematic diagrams will be necessary to effect these changes. If your receiver has a built-in audio limiting filter, this portion of the circuit may be changed in response by altering the filter or through bypassing it completely.

TRANSMITTING PROCEDURES

To make certain that a maximum amount of the transmitted signal reaches the receiver antenna, it is necessary to know a bit about transmitting procedures. If an external vertical antenna is used for transmitting, make certain that it is mounted in a

271

completely vertical manner. If mounted in a diagonal or horizontal configuration, some of the transmitted signal will be lost across the receiving antenna and into the ground.

Whenever possible, choose a remote control point which is free of large, metallic structures which can quickly dampen much of the output signal. Hopefully, this location will contain *no* metal objects, but if there are structures such as these present, keep them out of the line between the transmitting and receiving antennas. Line-of-sight should be obtained if possible. This means that the receive antenna can be seen from the transmitting antenna location. A clear path for the transmitted signal assures maximum strength at the receiver, as the transmitted signal is less affected by the distance between the two antennas than by obstructions which lie between them.

Make certain that the power supply to the transmitter is providing the correct voltage and is adequate to handle the current drain of the transmitter. Most of the time, batteries will be used to power the transmitter control. As these begin to weaken, the power and overall performance of the transmitter are affected. If the batteries seem to weaken quickly, some of the more expensive alkaline types may be more suitable or you may parallel two or more batteries to supply the same amount of voltage and current but for a period of time which is multiplied by the number of batteries in the parallel circuit.

If a transceiver or walkie-talkie is used for the remote control command source, make certain that you hold the unit so the antenna is as vertical as possible. Too often, the unit is tilted so the antenna is more diagonal or horizontal than vertical. Diagonal and horizontal transmitting antennas can be successfully used, but it is necessary to mount the receiving antenna in the same configuration to get the maximum transfer of signal between the two systems.

Inspect all antenna connections at the transmitter and at the receiver periodically for signs of lead breakage or rusting. This is especially true if the antennas are mounted out-of-doors where the elements can quickly take their toll on metallic structures and connections not adequately protected. Many types of weather-proofing compounds are available from hardware stores which will completely protect exposed metal surfaces from moisture. Salt is a problem around coastal areas and more frequent inspections of outdoor antennas will be necessary if you reside near the ocean.

The complexity of radio transmitter remote control systems is normally found in the method by which audio tones are superimposed upon the transmitter carrier, demodulated and acted upon in the receiver section. Getting these signals to the receiver and converting them back to audio tones is the simplest part of the process.

SUMMARY

Remote control over long distances can easily be accomplished by using inexpensive radio equipment available from most hobby stores and electronic supply houses. Little or no modification to the existing circuits is necessary in many applications. All that is required is to add a tone generator circuit for connection to the input of the transmitter and filter and switching system at the output of the receiver.

Inexpensive transceivers can be purchased and used about "as-is," or their circuits can be removed and mounted in a case which also houses the tone generator circuits and power amplifiers where legal and necessary. The same applies to the receiver unit which can be made from a similar transceiver circuit board and used for receive purposes only. In this latter case, the filtering and solid-state switches along with their relays can be housed in the same case.

Refer to the chapter on tone control, because most of the systems described were based on these former circuits. Radio control, in most instances, is tone control but using the transmitter carrier to enable the audio tones to be sent and received over a much longer distance and with no signals heard in the audio

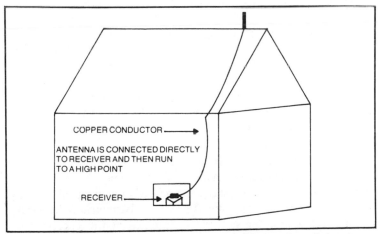

Fig. 10-11. Long-wire antenna which is connected directly to the radio receiver.

spectrum. Remote control is always a transference of energy from one system to another. In many of the circuits, methods, and projects discussed in this chapter, the audio frequency was then sent through space to be received by the detecting antenna. In the receiver circuitry, the original audio tones were, subsequently, fed through a filtering circuit to a solid-state switch. The actual transmitting and receiving functions were only a very small part of the overall system.

Pay close attention to the FCC rules and regulations, especially where they apply to licensing of transmitters and to the restrictions on power input and antenna height. These rules allow for very complex and long-range remote control applications. It is *never* necessary to exceed the limitations the FCC places on these circuits and devices for *any* remote control application. If your transmitter is not powerful enough to cover the distance between its antenna and the one at the receiver, the latter can be chosen to provide the sensitivity needed for proper control of the work load. Remote control can be accomplished over any reasonable distances by using circuits already available on the market. It will probably be less inexpensive, in the long run, to go this retail route, as the parts needed to build a comparable transmit-receive system would probably cost more than the simple, but effective units available at your local electronics outlets.

Finally, if you find it necessary to install outdoor antennas at the receiving and remote control sites, please pay heed to safety at all times. Each year, several experimenters are killed or seriously injured by their antennas coming in contact with voltage lines. Do not install *any* type of antenna in close proximity to an electric line, regardless of the voltage rating. Do not install antennas in such a manner that they cross over or under these same lines even if the clearance is perfectly adequate. Winds can make spaghetti out of electric wiring and antenna conductors, intertwining them so that energy from one system is directly transferred to the other. Leave room for all occurences. If your antenna site opens even a remote possibility of contact with other wires and cables, choose an alternate, safe location.

Lightning must also be considered when cables are strung to high points, far above the normal, surrounding objects. Make provisions for grounding your antennas whenever an electrical storm is forecast or threatening. A ground rod driven into the earth near a point where the antenna terminates at the receiver or transmitter is a convenient method of grounding. When the remote control circuits are not in use, an alligator clip lead can be used to connect the center portion of the coaxial cable to the ground stake. Antennas which use artificial ground systems such as radials mounted on a roof or pole should also tie the braid of the coaxial cable to the ground rod connection. Antennas which have good earth grounds in normal operation may simply contain a shorting switch at the base of the element near the grounding point. This switch will short the center conductor to the grounded braid and channel most of the static electricity built up in an electrical storm to ground. If possible, it is a good idea to get the coaxial feed line completely out of the building where the receiver or transmitter is located. This can be done by wrapping the termination end of the cable with stiff twine. When a storm threatens, the cable can be disconnected from the receiver or transmitter and dropped out of a convenient window to the ground. Later, when the storm has safely passed, the cable may be hauled in again by pulling in on the twine. This fishing technique is simple, costs almost nothing, and assures the safety of your home and your family during a severe electrical storm.

Whenever you work on transmitters and receivers, especially if they are of the older, vacuum tube designs, be on guard for voltage contacts which may present a hazard to you while testing the system initially with the protective covers removed. Solid-state circuits normally do not present this safety hazard, as the voltages are below the force level which is dangerous to humans. However, if AC line-derived power supplies are used, there is always a possibility of this 120-volt energy reaching the confines of your solid-state circuitry. This is a million to one chance and would take a massive, disastrous failure within the power supply circuitry, but it can happen. *Never* work on any piece of electronic equipment unless you know the circuit is not receiving and cannot receive power. Some adjustments must be made with the power supply operating and driving its load. In these cases, use extreme caution if significant voltages are present anywhere within the system, and never operate a circuit for any length of time nor allow it to be in a place where others may touch it unless it is safely installed in a metal or insulated case.

These safety tips may seem obvious, but each year persons are killed by forgetting some of the most basic procedures taught the newcomer to electronics experimentation. You only have to forget *one small rule* for a second or so, and you could easily be the next statistic in electronic accidents.

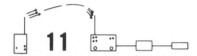

Carrier-Current Remote Control

Many of the systems for establishing remote control within the confines of a home or other building require that conductors be run from the remote control point to the sensors at the work load. Almost every building in America already has hard wiring run to every room and past every location in the form of the AC house current wiring which supplies 120 volts at each wall outlet. It seems a waste to have this wiring throughout the home and then to have to run other wires to about the same locations. Many of these electrical outlets closely border the area where the work is performed.

It is practical to use the hard wiring in your home for purposes other than what it was originally intended for and still preserve the original use. A remote control system which uses the same wires used to supply operating current for most household appliances is called a *carrier current system*. This is a form of hard wire radio transmission, because the activating tones are fed to the input of a transmitter. These tones are superimposed on the transmitter carrier as before, but in carrier-current systems, the transmitter feeds its output to the house wiring instead of sending the signals through open space. This can be made into one of the finest and most versatile control systems of all those discussed as long as the remote control arena remains within the home, proper. It can expand to the outdoors as long as the work load is located near an outdoor outlet to the AC power system.

There are quite a few problems involved in coupling a transmitter to the AC lines, because they were not intended to carry signals of a medium- to high-frequency nature. Normally, a transmitter frequency just below the broadcast band is chosen, because the AC line seems to adapt best as a conductor of these frequencies. The useful range of frequencies which can most easily be made to work lies in the radio spectrum between 100 and 450 kilohertz. Higher and lower frequencies may be used on occasion, but this will depend on the size of the AC line system and any spurious signals or noise that may be present within the system. Certain appliances such as mixers, blenders and other devices which contain AC motors can inject radio frequency signals into the AC line. This can cause interference in some remote control systems.

Figure 11-1 shows a block diagram of a carrier-current remote control system. It connects the transmitter to the line through a capacitive coupling circuit which assures that the AC current does not pass through the transmitter instead of the transmitted signal passing into the line. The signal is transmitted through the house wiring to the detection point where the receiver demodulates the signal and produces an audio output which corresponds to the audio input at the transmitter. From this point on, the system parallels others discussed in this text by passing the audio tones through appropriate filter circuits and on to the switching networks. The carrier current remote control system is exactly like the systems described in the previous chapter except for the frequencies used and the fact that hard wiring carries the signals to the receiver instead of them being broadcast through the air.

Figure 11-2 shows one of the easiest ways to get a carrier current system into operation. The device pictured is an inexpensive wireless intercom. Many people have seen or used these within the home, however most do not know how they work. This is a carrier current system which transmits voice modulation instead of tones. A microphone (which doubles as a loud speaker when in the receive mode) takes the input from the human voice and passes it on to the transmitter modulator circuitry. The signal is superimposed on the carrier as a varying, complex waveform. The receiver demodulates the transmitted signal and reproduces the human voice at its audio output.

For remote control purposes, a tone generator can be substituted for the microphone and the output of the generator fed directly into the transmitter. This can easily be done by removing the speaker leads and replacing them with an input transformer with an 8-ohm secondary to match the impedance normally presented by the small speaker. At the receiver end of the system, a filter circuit is connected to the audio output. Here, it may also be necessary to install a small output transformer to match the speaker output, or a microphone or other audio transducer can be used to pick up the audio tones. After filtering, the signal is fed to the solid-state switching circuits.

This is a very uncomplicated way of establishing carrier current control with a minimum of time and expenditure. In addition to the intercom units, all that is needed is a transformer or two and the tone generation and detecting circuits already discussed in earlier chapters. The commercial intercoms will do the rest after a few simple wiring and circuit changes.

THE HOME WIRING SYSTEM

There are certain problems in connecting a transmitter and receiver to the AC line, so the first examination should be at this line. It is necessary to take a thoughtful look at the system by which the radio frequency energy is to be transmitted to allow for transmitter frequency choices which remove the control command data from any interfering signals. An oscilloscope connected to the line will possibly provide some information, although the amplitude of the 60 hertz wave is so strong that weaker, interfering signals may be covered completely, but this is a good start. Many oscilloscopes will sample the AC line directly through internal connections through their power supplies. Look for any dim signals which may appear as noise or static on the cathode-ray tube, and try to determine what frequency or frequencies they are occurring at. Your scope may have the sweep range to lock into these spurious signals.

An easier method may be to connect two wireless intercoms to the circuit and listen to the output when one is locked into the transmitting mode without receiving any audio input. It will be necessary to know what frequency or frequencies these intercoms are transmitting, and receiving on. This information can usually be obtained from the owners manual and schematic drawing if one is included with the sets. An oscilloscope can be connected to the receiver output for a better look at what's taking place on the line.

If the received signal from the intercom is clear and relatively free of hum and noise, this frequency would be a good start to attempt remote control operations. If a homebuilt transmitter and receiver are to be used, try designing their circuits for the same frequency as that of the intercoms. Homebuilt circuits using

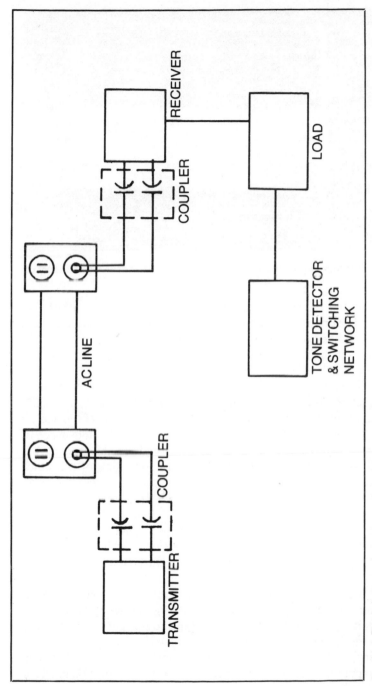

Fig. 11-1. Block diagram of a carrier-current system.

279

Fig. 11-2. Inexpensive wireless intercom can be used for carrier-current control.

integrated circuits are quite common and can be easily adjusted to operate across a broad frequency spectrum below the broadcast band. If none of these testing procedures is practical at your site and a homebuilt transmitter is to be used, try a beginning frequency of 100 kilohertz and work from there.

Devices which are connected to the AC line will have a direct impact on the operation of the carrier current transmitter. Circuits such as electrical lights will provide a path for some of the energy put in the line by the transmitter. Fortunately, the impedance of these resistive devices is moderately high at the frequencies normally used for carrier current operation. Motors and other devices which are inductive in nature due to their windings may cause many more problems, not because of their inductance but due to distributed capacitance between windings. Because of the close spacing of most of these copper windings, significant capacitance can be present in these devices which may have a swamping effect on the carrier current signal. Figure 11-3 demonstrates this last condition.

One of the easiest remedies for the above condition is to remove the devices and appliances causing the problems from the circuit. An alternate house circuit (on another buss) may provide enough isolation to prevent the absorption of the carrier current energy. Alternately, an inductive trap may be installed between the appliance and the ungrounded side of the AC line. Figure 11-4 shows the trap which is easily and inexpensively made if the appliance it is to be used with does not draw a lot of current. If it does, a very large inductor must be wound with heavy copper conductor. The complexities mount here, as the size of the inductor makes it difficult to safely house the circuit in a protective, grounded aluminum box.

Another solution to a severe swamping problem is to make the carrier current transmitter variable in frequency. The output frequency can then be adjusted over a fairly wide range, choosing the frequency which provides the greatest amount of signal input to the receiver which must also be made with a variable frequency control.

COUPLING THE TRANSMITTER AND RECEIVER TO THE LINE

It is necessary to adequately design a coupling network which will allow the rf energy from the transmitter to be transferred safely to the AC line without allowing any of the AC current to pass back through the transmitter. Fortunately, coupling networks are not complicated and can be built in a short time period for little expense. What is vitally important in their design and construction is that the circuits be completely enclosed in grounded metal boxes to prevent accidental contact and possible shock hazards. The metal box should be directly grounded to the AC ground source which can be had directly at a three-wire outlet. Figure 11-5 shows a method where the coupling network and its metal enclosure can be constructed so that the entire unit simply plugs into one of the outlets. The connections to the AC line are automatically made, and the box grounded at the same time.

The circuit consists simply of two 0.1-microfarad capacitors. Each is connected in series with the transmitter output or receiver input and the AC line. The capacitor value is not critical and any capacitors within the range of 0.1-microfarad will work well. The idea here is to provide an input path from the transmitter to the line which is one way only. While signals from the transmitter may travel into the line, most of the AC energy cannot pass back

Fig. 11-3. Illustration of signal swamping effect brought on by the distributed capacitance between the turns of a coil winding.

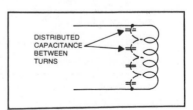

DISTRIBUTED CAPACITANCE BETWEEN TURNS

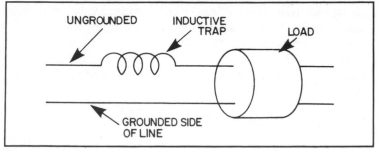

Fig. 11-4. Inductive trap installed between the line and various devices may prevent severe swamping of the carrier-current signals.

through this coupling circuit. The metal housing is attached directly to the third pin of the line plug and to the AC ground through this connection. The voltage ratings of the capacitors should be at least 1000 volts DC and higher values are perfectly acceptable. Do not use capacitors of lower ratings, as they may fail due to voltage spikes. A shorted capacitor is a rare occurrence, but if it should happen in this circuit, the results could destroy the output network of the transmitter and present a safety hazard.

The circuit shown in the last drawing is a direct method of AC line coupling. There are alternate methods of getting the transmitter signal onto the line that are less complicated. The results using the system of Fig. 11-6 will vary depending on the condition of the AC line. This latter circuit inductively couples the output from the transmitter to the line. No direct line contact is made, so the safety of this circuit is maximized. It consists of a long extension cord wound around a slender wooden dowel rod. The output from the transmitter is also fed into a similar extension cord, although this can be a length of zip cord without the plug and receptical at each

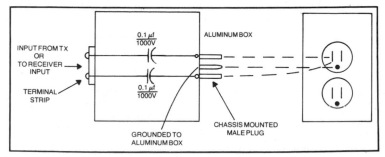

Fig. 11-5. Circuit for coupling carrier-current transmitter or receiver to the AC line.

end. The idea is to intertwine the cords on the coil form. To do this the two cables are placed side by side and then simultaneously wound around the one-half-inch dowel rod. The longer cord lengths are to be preferred as they provide better coupling. When the coil is wound, it can be secured by electrical tape. The male plug from the first extension cord is then inserted in the wall outlet. The energy from the transmitter will be transferred to the first extension cord which is directly attached to the AC line. It is a good idea to apply electrical tape windings around the entire length of the coil. This will keep the windings secure and will provide additional insulation protection. Make certain that both cords are new and that there are no nicks or cuts in the insulation of either. Locate the coil near the AC outlet for connection purposes, and try to keep it clear of any large metal objects which will serve to detune the system. Total coil length should be about one foot, although the longer cords may necessitate a slightly larger winding.

The inductive coupling system under discussion may be tuned slightly by wrapping a strip of tin foil around the coil. This should be no more than four inches in diameter. Figure 11-7 shows how this is attached. The foil is would loosely around the coil. Tuning is accomplished by moving the foil across the coil while observing the received input from the transmitter. When the signal is at a maximum, the foil may be permanently secured to the coil with more electrical tape.

TRANSMITTING FILTERS

Even though the coupling capacitors block most of the AC energy, a small amount of 60-hertz energy may still travel back through the transmitter circuitry. This can cause hum problems

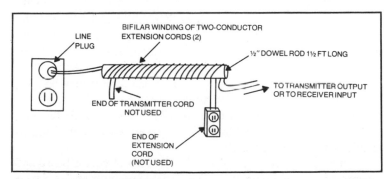

Fig. 11-6. Capacitive and inductive method of coupling signals to the AC line while requiring no direct connections.

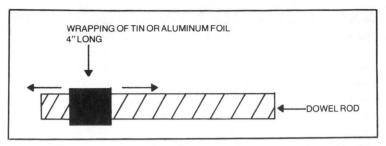

Fig. 11-7. Tuning arrangement for inductive coupling systems using a strip of tin foil.

within the electronic circuitry which is heard as noise at the receiver. This same 60-hertz hum can travel through the coupling network of the receiver and will be heard as noise or hum at this point of the control system. To avoid these problems, it may be necessary to install 60-hertz line filters at the transmitter and at the receiver if hum is being experienced. These filters are easy to build with standard components, although, it may be necessary to wire some resistors in series to arrive at the desired values. Alternately, variable resistors can be used which will allow the fine tuning of the system to prevent AC hum.

Figure 11-8 shows the circuit which is coupled to the output of the transmitter and feeds the capacitive coupling network discussed earlier. It uses variable resistors and fixed capacitors to pass the transmitter energy while blocking the 60-hertz energy to a high degree.

This is a twin-T network similar in many ways to the circuit used in the five-tone generator project of an earlier chapter. The values have been changed in this latter circuit to block the low frequency energy.

Most types of capacitors will work in this circuit, although the electrolytic types are not applicable. Surplus, oil-filled capacitors may be purchased inexpensively or common disc ceramic types may be used. The two smaller capacitors will be easily found in disc ceramic components. The larger, 1-microfarad unit may be a bit more difficult to come by, and an oil-filled type may be required. Voltage ratings are not critical as was the case with the line coupler, but values of 1000 volts or more are easily obtained and should be used when possible.

This circuit can be built on a small piece of vector board and mounted in an aluminum case. The variable resistor shafts can be brought through the case for constant adjustment availability or can be mounted on the circuit board, adjusted and then sealed in the metal case.

Using an ohmmeter, set the variable resistors, R1 and R2, to a value of 2688 ohms. It will be necessary to use an accurate, digital ohmmeter to set these controls this closely, but if one is not available, set them as closely as possible with a standard ohmmeter. Later, the components can be fine-tuned to filter out the undesirable hum by listening to the receiver. Resistor R3 is also measured on the ohmmeter and set to a resistance of 2688 ohms before wiring it into the circuit. After the wiring is completed, measure these values again to make certain they have not fallen out of adjustment. Here, 5000-ohm controls were chosen for R1 and R2, because the desired setting of each unit will place it at the center of its resistance range. A 10,000-ohm control is used for R3 for the same reason.

If the resistors were correctly set before sealing the box, no further adjustment should be necessary, although you may want to vary R3 slightly to see if a slight improvement can be obtained. If hum still persists, adjust all controls a 32nd of an inch each time. Through a little experimentation, a decrease in hum level should be obtained.

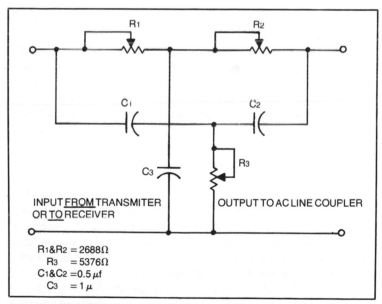

Fig. 11-8. Filter circuit which is attached between the carrier-current transmitter and receiver and the line coupling circuit.

This circuit is identical for both the transmitter and receiver. It may be necessary to use a filter at both units, although, in many instances, one is only needed at the receiver. You might start by building one unit and testing it with the receiver alone. If this does not reject the hum, place the circuit at the transmitter. If the hum still persists, build an identical filter circuit and place one at both the transmitter and the receiver.

Other values of components may be used for this circuit if they are more readily obtainable than the ones specified. C1 and C2 are always identical and must be half the capacitance value of C3. R1 and R2 are also identical and are each half the value of R3. If the capacitance of C3 is doubled, then C1 and C2 are also doubled. R3 is halved in value as are R1 and R2. By remembering these ratios, a wide range of components may be used to build this simple circuit. If C3 is halved, R3 will be doubled as will the values of R1 and R2. C1 and C2 will be halved in value.

Figure 11-8 shows a block diagram of the placement of the filter in the circuit between the line coupler and the receiver or transmitter. It is desirable to separate the filter circuit from those of the receiver or transmitter by mounting the filter in a separate case. The coupler, however, can be built on the same vector board of this latter circuit.

THE CARRIER TRANSMITTER

Depending on the desired frequency, many different circuits can be made to serve as carrier current transmitters. Many of the wireless broadcaster circuits seen in electronics projects books can be made to work. These often transmit on the AM broadcast band but can be tuned to frequencies below the low end of these commercial frequencies. Figure 11-10 shows one of these simple circuits which uses a single transistor and a ferrite rod loopstick antenna available at most hobby outlets. The circuit is constructed on a piece of perforated circuit board. Keep all wiring as short as possible and separate the coil portion of the circuit from the tone generator input portion by a few inches. This oscillator is tuned by adjusting the variable capacitor in parallel with the loopstick antenna. The tuning is rather critical, and it will be necessary to connect the transmitter to the line and set the desired frequency by feeding a tone to the transmitter input and listening for it at the receiver which has already been set to frequency.

This simple circuit may be adequate to provide acceptable remote control in non-critical applications, but the transmitter will

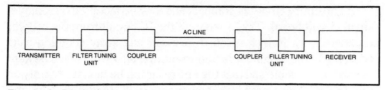

Fig. 11-9. Block diagram of complete carrier-current system showing the placement of various circuits.

tend to drift off frequency when various appliances are attached to the AC line. A tuning network previously described may provide some isolation from the line and is desirable, at least at the transmitter output.

A more desirable transmitter circuit is shown in Fig. 11-11 and it is not as complicated nor as critical in responding to varying line conditions. It uses an integrated circuit and a few other components and may be readily varied in output frequency to enable the user to select the one least affected by line noise.

In addition to the integrated circuit and its associated components, there is a two-transistor output stage which serves to isolate the frequency controlling circuitry from the line load. This latter circuit is called a *Darlington pair* and serves to transform the impedance of the frequency controlling circuitry to more closely match the line load which is extremely low. This circuit acts as a *buffer* stage to provide good isolation between the transmitter and the line, so line load variations will not tend to pull the frequency as much.

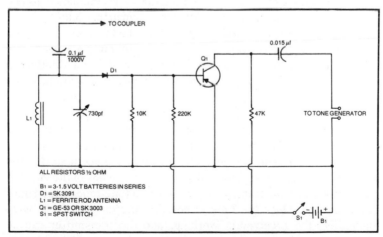

Fig. 11-10. Carrier-current transmitter circuit composed of a one transistor oscillator.

The entire circuit is compact and may be mounted on a single, small circuit board. Try to separate the IC circuit from the coupler transistors by an inch or so to prevent any stray coupling from occuring in the circuit. It is important that all wiring leads be kept as short as possible and that the entire circuit be mounted firmly on the board. Movement of components will tend to make the frequency shift. After the circuit is completed and known to be working properly, a small drop of the epoxy cement will secure all of the components to the circuit board and prevent their moving due to mild shocks.

This circuit can be varied over a very great frequency range of from about 150,000 hertz to about 1800 hertz. The lower end of this frequency spectrum is back in the audio range again and is not recommended; 50,000 hertz is about as low as most applications permit. This allows a lot of room for selecting the various control tones which lie in the audio range and are broadcast over the carrier of the transmitter through the AC line.

RECEIVER CIRCUITRY

The receiver is an all-important part of this system, because it detects the audio tones by demodulating the carrier. Fortunately, these can also be simple circuits which provide acceptable operation in carrier current systems without costing more than five or ten dollars to complete.

An extremely simple design is shown in Fig. 11-12 which uses a single transistor. Notice the remarkable resemblance to the one-transistor transmitter recently discussed in this chapter.

The signal is detected across a small diode which is attached directly to the line coupler and filter tuning unit if one is used. The diminutive output from the diode is then amplified and fed to an audio amplifier. The output is sufficient in most applications to not require a preamplifier in the output stages of the detector.

Wiring should be kept as short as possible, and the detector circuit consisting of the coil, variable capacitor and diode should be separated an inch or so from the transistor amplifier circuitry.

After the circuit has been completed on a piece of perforated circuit board, it can be activated and tuned by means of the variable capacitor while the transmitter is feeding a tone-modulated signal into the line. Once the tone is heard at the output of the audio amplifier, the receiver is working, and the transmitter can be slowly moved in frequency if desired to reach a more favorable part of the spectrum which is immune to noise. This simple receive

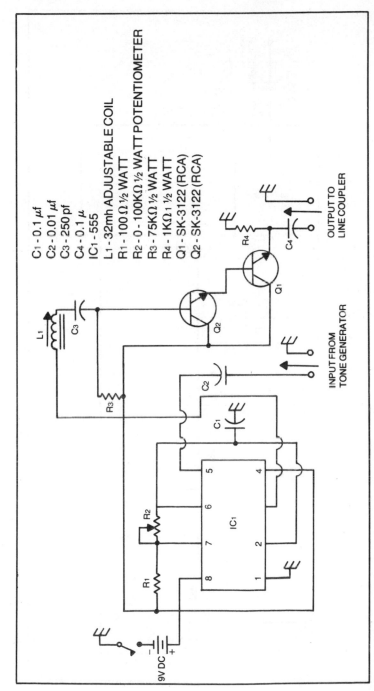

C₁ - 0.1 μf
C₂ - 0.01 μf
C₃ - 250 pf
C₄ - 0.1 μ
IC₁ - 555
L₁ - 32mh ADJUSTABLE COIL
R₁ - 100 Ω ½ WATT
R₂ - 0 - 100KΩ ½ WATT POTENTIOMETER
R₃ - 75KΩ ½ WATT
R₄ - 1KΩ ½ WATT
Q₁ - SK-3122 (RCA)
Q₂ - SK-3122 (RCA)

OUTPUT TO
LINE COUPLER

INPUT FROM
TONE GENERATOR

9V DC

Fig. 11-11. Integrated circuit carrier current transmitter.

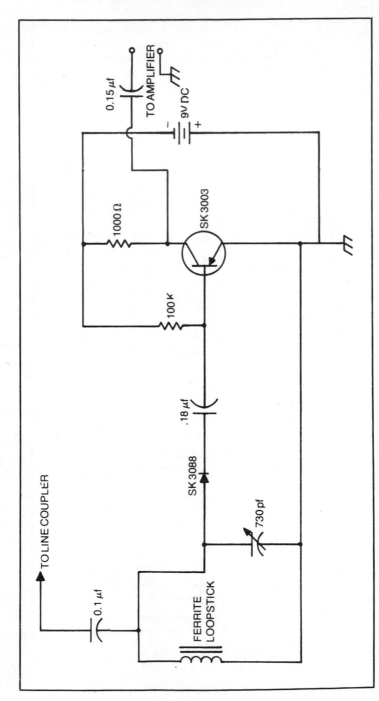

Fig. 11-12. Single transistor carrier current receiver circuit.

circuit should provide reasonable stability at least matching that of most transmitters used for these applications.

Another receiver circuit shown in Fig. 11-13 uses the type 557 integrated circuit. Closely resembling the previous transmitter circuit which used an integrated circuit, this receiver is variable over a wide frequency range.

Construction is handled in much the same manner as with the transmitter circuit, and wiring connections should be kept as short as possible. The variable resistor controls the frequency range and is used to tune the receiver to the transmitter frequency. In choosing R1, make certain it is a linear control. Some potentiometers have what is known as an *audio taper* which does not provide smooth adjustment in this type of circuit.

The output of the receiver may be adequate to directly drive the filtering and switching circuits of later stages. If not, an audio amplifier will have to be inserted in the output line between the switching and filtering portions of the circuitry.

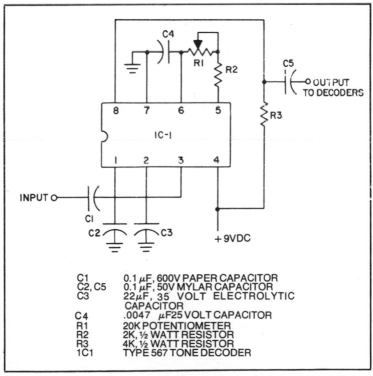

C1	0.1 μF, 600V PAPER CAPACITOR
C2, C5	0.1 μF, 50V MYLAR CAPACITOR
C3	22μF, 35 VOLT ELECTROLYTIC CAPACITOR
C4	.0047 μF 25 VOLT CAPACITOR
R1	20K POTENTIOMETER
R2	2K, ½ WATT RESISTOR
R3	4K, ½ WATT RESISTOR
1C1	TYPE 567 TONE DECODER

Fig. 11-13. Carrier-current receiver using integrated circuit design.

SUMMARY

A lot of experimentation is usually required both at the transmitter and at the receiver to get carrier current systems to properly operate. The main problem is encountered in matching the units to the AC line in such a way that a fairly efficient transfer of power is obtained. Line noises and loads will also have to be discovered, and the frequency of operation and the filtering circuits designed to overcome these problems.

Once the system is operational, it should provide dependable remote control operation as long as the AC line load does not change drastically. Extremely narrow filter passbands may be needed at the switching portion of the circuit to prevent false triggering of the work load if especially noisy conditions are present on the line, but for most applications, the carrier current system offers a viable alternative to separate hard-wiring of the remote control command signal lines.

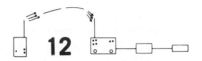

12

Obtaining Components

Once a remote control system has been designed on paper, it becomes necessary to locate the components which are needed to build the system. Some of these items may be very difficult to find or too expensive, necessitating changes in the original design. It is best to design the ideal system and then make a diligent component search. Make a list of the components which are readily obtainable, those that are not, and those which can be had but at an objectionable price. Now, go back and look your system over again and figure out modifications which can use other circuits or components to replace those which cannot be obtained. Also, check to see if the expensive components might be replaced with alternate circuits and designs. On this second design check, you will have a much better idea of what is available and reasonable and be able to intelligently re-design the original system to arrive at one which performs the desired functions and is practical to construct.

The electronic parts usually do not present much of a problem, as most hobby stores and electronic outlets will carry the various capacitors, resistors, switches, relays, etc. Some of the transistors and integrated circuits may need to be ordered, but before doing this, check to see if a direct replacement component is available from another manufacturer. Most transistors and integrated circuits carry the manufacturer's labeling which is like their brand name. For instance, an RCA transistor labeled SK 3012 is the same as a generic 2N2144. This same transistor is known as a GE-16 when purchased from General Electric.

Fortunately, most electronic outlets carry cross-reference data which will quickly allow them to give you a direct-replacement component if you provide them with the specified number or identification of the part shown in a schematic diagram. Some of these devices will not *look* exactly like the parts indicated, if pictured, and the lead configurations may be different from the indicated device, but all direct-replacements can be made to work in most circuits. All manufacturers identify the lead configurations of their devices on the packing carton or in a technical data sheet supplied with each component.

For some of the heavier and exotic relays, it may be necessary to go to an electrical supply house which stocks these parts for control of high-current loads such as motors and industrial heaters. Just about any switching combination is available, but most of these parts are relatively expensive. A good alternate solution is to subscribe to the catalogues of the various surplus supply houses which buy up industrial and government surplus parts in large quantities and at extremely low prices. These prices are, of course, passed along to the consumer. For example, the author obtained a government surplus relay through one of these outlets which was electrically identical to a unit being sold through an electrical outlet. From the surplus business, the total price (delivered) was $2.95. From the electrical outlet, the price was $78.90—quite a savings!

Those parts which are bought on the surplus market may be used merchandise, but most function properly and are made to last. While guarantees vary, most reputable dealers will consider replacing those parts which do not prove satisfactory within a 30-day period. If worse comes to worse, and a component fails, another can be ordered, and you have still saved by several hundred per cent over the price of a new, commercial replacement.

Many of the circuits you plan to build may already be constructed and available at very reasonable prices from the surplus dealers. They may require slight modifications, but they are probably of very high quality and may provide a lifetime of trouble-free service. It can cost a little more than the original purchase price to modify these circuits and mechanical devices, but the price may be far less than constructing the same devices from scratch by buying, retail, all of the necessary parts.

A good rule in constructing remote control systems is to constantly read the available catalogs from all sources. Some of the items pictured and displayed may lead you to think of ways of incorporating them into a system. You may perform, in a single, compact unit, the functions of several discrete circuits and mechanical devices. The cost may even be less than all of these discrete parts combined.

One market the author taps quite often is the electronic toys industry which has come out in recent years with a tremendous variety of remote controlled toys which can perform one or many remote functions. For instance, one manufacturer's remote controlled robot was purchased in a damaged state from a department store for five dollars. It cost nearly $40 retail. The model had been dropped on the floor and the plastic case was badly cracked, however the electronic circuitry was completely undamaged, and a complete two channel, radio-controlled remote system was salvaged which boasted two separately controlled motors and slow-speed gearing trains. The sensing circuits were controlled by a two-tone radio transmitter which easily fit in the palm of the hand. The circuit was eventually used to control a remote shooting range target system which, by command, could set up a complex series of target movements. Additionally, the pierced targets could be motored to the shooting location by means of a string and pulley clothesline arrangement which was powered by one of the two motors in the toy circuit shown in Fig. 12-1.

This is just one example of what can be had if you look at various products with remote control applications in mind. The toy under discussion has very complex circuitry and mechanics. The electronics portion is wired on a tiny printed circuit board and can fit inside a pack of cigarettes.

While a circuit such as this could be built at home, the physical layout of components, and a sizable expense would be incurred in purchasing the compounds and blank boards necessary to produce a finished product in a single unit quantity.

Figure 12-2 shows two of the remote controlled automobiles sold by Radio Shack. Each is controlled by a radio transmitter which transmits tones to a tiny receiver within the plastic frame of the model. An electric motor powers the cars, and small solenoids are activated to freeze one of the front wheels to effect turning. The power to this solenoid could just as easily be used to drive a small relay which, in turn, could switch a larger relay and activate a high-current work load. These products have a limited range, but they are very satisfactory for control within the rooms of a small home. Some may even be able to trigger remote systems from the back yard.

Fig. 12-1. Versatile, multifunction remote control circuitry removed from children's toys.

Another Radio Shack product is shown in Fig. 12-3. This is a motor-driven automobile antenna which raises and lowers the element upon switching 12 volts DC to the motor input. The gearing mechanism allows for a fairly slow rpm at the output and the motor is reversible. Normally selling in the forty dollar range, this item had been dropped and the antenna broken. It was offered for sale on an "as is" basis for $9.95. For remote control purposes, the antenna was not needed, and the motor was used to open and close the drapes in a home. Power was supplied to the motor from a transformer and rectifier circuit that produced 12 volts DC from the AC line. The system was activated by radio control from a circuit and transmitter taken from a remote controlled model car. Total cost of the entire control system was less than $22. To build a system such as this from new parts might cost several hundred dollars or more.

Another excellent source of inexpensive electronic circuits for remote control applications is through kits which are sold at many hobby stores. These provide all of the parts needed to build a specified circuit. They also provide complete building and testing instructions. Troubleshooting information is even available if the circuit doesn't work properly on the first testing. Depending on the circuit, these kits can save the builder from ten to over 50 per cent on the cost of a pre-assembled version.

The best known manufacturer and distributor of electronic kits in the United States is Heath. Much of the test equipment in this book was built from their kits which have a reputation for being easy to construct with the step-by-step instructions provided. They also have an excellent technical staff which is available by phone to talk the builder in on a problem should the trouble-shooting instructions provided with each kit fail to bring the problem to light. Heathkit is of special interest to the remote control enthusiast who may desire to assemble some of the components for his system through a commercial kit. This company offers several remote control transmitter/receiver kits which are designed to be used with model airplanes for remote flight control. Presently, they offer transmitters with up to eight discrete channels of operation. These are complete, when

Fig. 12-2. Radio Shack remote controlled automobiles and other devices contain valuable circuitry for remote control applications.

Fig. 12-3. Radio Shack motorized antenna offers many parts for remote control applications.

finished, and offer the switching and channel select functions on the master control units.

Additionally, they offer sub-miniature servo units which produce mechanical motions of several types. Again, these kits are designed for remote control of gas-operated model airplanes, but they can easily be modified to trigger remote relays, solenoids, motors, etc. If you don't wish to go for the entire, in-case, package, Heath also offers transmitter and receiver modules which can be custom-wired into alternate circuits of the builders choice. These modules are available for all of the remote control channels from 26 MHz to 76 MHz. Figure 12-4 provides a table of specifications for their top-of-the line remote control transmitter which has a 500 milliwatt input. Each is extremely light in weight and easily handled as a remote unit in even the tightest places where miniaturization is a must. Table 12-1 provides a specifications breakdown on their servo units which provide power to the mechanical flight controls. It will take some ingenuity and experimentation on the part of the builder who wishes to use these units for other than control of model airplanes, but amazing results can be obtained when the principles of force and movement discussed in earlier chapters of this book are applied to these circuits and modules.

MECHANICAL PARTS

One of the finest sources of mechanical parts such as switches, relays, pneumatic controls, low-speed motors, etc. is probably sitting in your garage or driveway at this moment. The automobile provides a wealth of these parts, and some of the later models even have pneumatic controls which operate from air pressure.

A wrecked automobile can often be had for the price of towing it away, depending on the amount of damage. Even one that has been termed a total loss will still yield most of the parts of interest to remote control enthusiasts in perfect working order. Windshield wiper mechanisms and motors can be modified to perform a myriad of mechanical work functions. Electric seat motors produce an extremely high torque, and the control circuitry is already provided. All you have to do is remove these items from the chassis.

Similar in operation to hydraulic controls, the pneumatic types are often used to select the desired venting operation for air conditioning and heating systems. Instead of cables or hydraulic lines, these controls feature pneumatic tubes which sample the air from the manifold to obtain pressurization. A pneumatic valve channels the air to various portions of the system. Looking closely, you will find pneumatic valves, actuators, and many more mechani-

---

GDA 1205 and GDA 1919 1 SERIES SPECIFICATIONS: TRANS.—RF Frequency: crystal controlled, all bands. **Frequency stability:** within ±.005% on 27 MHz; within ±.002% on 53 MHz; 72 MHz. **Temp. Range:** 0 to 160°F. **RF Output Circuit:** pi network. **RF Input Power:** at least 500 mW. **Modulation:** on-off carrier keying. **Approx. Current Drain:** 100 mA on all bands. **Controls:** GDA-1205, 8 channels (4 with trim), on/off, trainer button; GDA-505, 5 channels (4 with trim), on/off switch, trainer button. **Power Supply:** Internal 9.6V 500 mAH Ni-Cad. battery. Rechargeable simultaneously with receiver battery at 35 to 40 mA from 120 volt power line. **Dimensions:** 7″ H x 7″ W x 2″ D. **Net Weight:** 2½ lbs., with battery. **GDA-1205-2 SPECIFICATIONS: RECEIVER—Local Oscillator Operating Frequency:** crystal-controlled, all bands. **Temperature Range:** 0 to +160°F. **Sensitivity:** 5μV or better. **Selectivity:** 6 dB down at ±4 kHz; 30 dB down at ±9 kHz. **Current Drain:** 10 mA. **Intermediate Frequency:** 453 kHz. **Power Supply:** 4.8V battery pack (GDA-1205-3). **Controls:** On/off switch. **Dimensions:** ⅝″ H x 1¾″ W x 1¾″ D. **Net Weight:** 2 ozs.

Fig. 12-4. Specifications for Heath remote control transmitter.

cal linkages which can be directly applied to remote control operations.

Every automobile has switches which perform varied functions. The headlight control, for example, is often a switch and a potentiometer or rheostat combined. An outward motion of the switch turns the lights on, while a clockwise rotation of the same control makes the dash lights brighter. A switch which operates on an "in-out" sequence as does the headlight switch can be connected directly to a solenoid. This latter device is found in abundance in most modern automobiles. The starter solenoid produces a large amount of force and will handle many heavy, short-travel, work loads. Relays are also very common. Some control high current devices while others are for smaller power systems.

Other miscellaneous items which will come in handy include the various lightbulbs and bulb sockets which can aid the construction of a light-activated remote control system. Fuse blocks of fairly complex design are found in most modern vehicles and can be used as the central buss of the power system. Blinking relays, some of them solid-state, can also be used in triggering stepping switches and in automation applications. Autos with electric windows will have another set of motors and a number of momentary switches for controlling them in either direction. Torque here is good, and speed is slow which make these devices ideal for remote control applications in motor-driven power systems.

HYDRAULICS

Many vehicles boast pop-open trunk lids. Sometimes, this is handled by solenoids but often, hydraulic actuators do the job. These are often sizable and able to exert a great deal of lifting force. Convertibles are just about a thing of the past in the United States which means there are probably many rusting away in junk yards. Hydraulic systems often control the automatic top lowering and raising procedures. The control system is within the automobile, so instead of just parts, here is an entire hydraulic system which will need little modification to operate in a remote control application. The luxury automobiles may have even more hydraulic equipment. Some of the newer models will even boast onboard computers.

Before you go out and spend several hundred dollars on parts for a fairly elaborate remote control system, look over the selection at your local junk yard. Check with an insurance company and find out how to go about making bids on their wrecked automobiles which are considered total losses. You may even be able to make a deal with a local garage to take a body off their hands after they have removed the engine, transmission and the rest of the drive chain which is really of little or no use in remote control applications. A wrecked, unrepairable automobile is simply junk to most people, but to the remote control enthusiast, it can be a gold mine, supplying working or repairable parts which might cost thousands of dollars if purchased separately through standard channels.

APPLIANCES

Progress is being made in most consumer-oriented appliances. So much so that many boast a wealth of electronic, electrical, and mechanical parts and devices. These are good hunting places for remote control system components. For example, a microwave oven which has a burned out magnetron tube may be discarded as junk, but if you search the internal circuitry, you will find a high voltage power supply capable of about 1000 watts of power input. You will also find a number of interlocks (used on the oven door), switches, solid-state controls and even a motor or two. Some of the more modern versions will have the famous "touchmatic" control panels which are solid-state in design and perform switching functions without physical movement. Some even have a solid-state clock which can be used to

Table 12-1. Specifications for Heath Servo Units that Provide Power to the Mechanical Control Devices.

SERVO COMPARISION CHART	Min. 1205-4	Sub-Min. 1205-5	Hi-Torque 1205-8
Idle Current (max.)	15 mA	15 mA	20 mA
No Load Current	80 mA	80 mA	150 mA
Stall Current (nom.)	450 mA	450 mA	1000 mA
Thrust	4 lb.	4 lb.	6 lb.
Travel Time	0.5 sec	0.5 sec	0.4 sec
Rotary Travel	90°	90°	90° or 180°
Dimensions (HxWxL*)	1⅝x15/16x2 9/32″	1 11/16x ¾ ×2	15/16x2 9/32″
Net Weight	1.75 oz.	1.25 oz.	1.75 oz.

All servos: **Power Requirement:** 4.8 VDC. **Position Accuracy:** 1°. **Mechanical Output:** 1 rotary arm, 1 rotary wheel. **Input Signal:** Pulse, 1-2 msec, 4V P-P. **Temperature Range:** 0° to 160° F. **Backlash:** <0.002″. *H includes outputs, L includes mounting ears.

automatically key certain commands into the remote system at preset times.

Conventional ovens may also have solid-state control systems in addition to heat sensitive switches and circuit breakers. The newer models have digital clocks like the microwave oven while the older units display time by the more conventional electric clock. Either can be used for timing applications in remote control systems.

Refrigerators also have temperature sensitive switches and relays. Some have a fair amount of plumbing which may be used in simple hydraulic applications, and, of course, there's the compressor motor which may be used as the mechanical drive for a hydraulic actuator with the proper gearing or hydraulic pump. You will also find a number of holding switches used to switch the light on or off as the door is opened and closed. The thermostats can even be put to use in certain remote systems.

TELEVISION RECEIVERS AND STEREOS

For the builder who has the patience to salvage parts from previously built circuits by desoldering, television consoles and stereos offer a virtual motherlode of components and whole circuits. Remote controlled televisions have been on the market long enough now, and many have long since passed their days of usefulness for their originally intended purposes. A color console with a bad picture tube and crumbling high-voltage section may be discarded as junk, because it is no longer practical to return it to a state of operation, but many of the other circuits may be in nearly perfect shape. Remote channel change sets offer a small motor which is ganged directly to the tuner in many of the older models. This was before the days of varactor tuning which did away with many of these slow-turning motorized activators. Often the remote ultrasonic tone generator is saved for parts, but these can be had for a few dollars from many television outlets who are swamped with pull-out parts. It is conceivable that a junked television might offer all that was needed for a three or four channel ultrasound remote control system.

Televisions will also offer a wide range of transistors, switches, potentiometers, and variable inductors and capacitors for making tone filters. Transformers are available in these models which can be made to provide operating power for many kinds of electronic circuits. Some of the later models may even have their circuits constructed on separate printed circuit boards which make initial removal from the chassis simple. Some circuits, if operating properly, may even be used with little modification as amplifiers and preamplifiers.

Stereo systems will offer many of the same electronic components of the television receiver. Some will have motorized tuning, although these will be the exception rather than the rule. Audio power transistors can be used for electronic switching of medium current loads, and the assortment of potentiometers and switches can always be used to good advantage by the enterprising remote control enthusiast. Even the manual tuning models with string driven tuning mechanisms can be put to use in certain systems. There are various control lights, and some of the later models will use light-emitting diodes. Those with digital frequency readouts are even more versatile for remote control applications, but these are few and far between on the junk market.

Repairable AM radios can often be slightly modified to make excellent carrier current receivers, while the FM receivers can be converted to receive the 76-MHz remote control channels and some will even tune down to the 50-MHz channels for amateur radio operator control. The ferrite loopstick antennas found in miniature transistor AM receivers can be used to build carrier current transmitters, and the miniature matching transformers used to feed the output of a tone generator or receiver to an SCR switching circuit.

WAR SURPLUS

Many of the surplus outlets in the United States specialize in materials from as far back as World War II. Much of this equipment is in excellent shape and can be had for a fraction of a cent on the dollar it cost to develop and manufacture them. Since the mid-1970s, many highly complex pieces of communications equipment have appeared on the surplus market. Some are unusable for their original design purposes but have a wealth of cams and gears, and small motors used as the mechanical power sources. Old radio-teletype machines are highly complex devices and nearly 100 percent mechanical in design. Many human mechanical motions can be closely simulated using the mechanisms from these units.

Relays and large solenoids will also be found in some of the larger pieces of transmitting equipment. Because of their size, they are often referred to as "boat anchors" meaning that all they are good for is to be thrown in the water as dead weight. The disassembly process may be a long and complicated one, but the author still has useful parts from an old transceiver taken apart over 10 years ago. Many of the components have been incorporated into electronic and remote control equipment in use today. The cost of obtaining these parts from a standard outlet (if they could be obtained at all) would have run into the thousands of dollars. The unit was given away at no cost by the dealer. All the author had to do was to haul the "eyesore" out of his store. This deal was not highly unusual, and many times, "worthless" pieces of equipment are more of a headache than a godsend to busy surplus dealers who need storage space. All that is required to acquire much of this equipment is the loan of a truck and a strong back (and stomach, sometimes).

LOCAL UTILITY COMPANIES

One important source of, often free, electronic and mechanical components is your local electric and phone companies. During the writing of this text, the author was given a complete carrier current transmitter/receiver by the local electric utility. It was an old, tube-type unit installed in a five foot steel cabinet. It was to be hauled away as junk. While heavy and huge, it performed to power company specifications upon activation and supplied a tone-modulated carrier on 81 kHz that could be switched to transmit and receive voice modulation.

On past occasions, touch-tone telephones, teletypewriters, and various assortments of latching relays, stepping switches, solid-state controls, etc., were dumped by the phone companies. By checking with the proper authorities, it may be very possible to have your name placed on a list to be notified when these dumpings are about to take place. All you have to do is haul this equipment away. It may have originally cost in the tens of thousands of dollars, but it's now considered junk, because it is obsolete or beyond the state of practical repair. A useless telephone may still contain an operational touchtone pad which will cost about forty to fifty dollars on the retail electronics market. Some of the switching equipment will contain audio tone filters which can be directly adapted to touchtone remote control. Miniature switches and relays are also in abundance.

If you get your name in early, you may be able to take advantage of these giveaways which will save you many expenditures in the future. The utilities are usually receptive to these transactions, because they save them time, trouble and the expense of having their equipment hauled to a dumping site.

SUMMARY

Always keep your eye to the ground, so to speak, by being on top of the surplus market at all times. Catalogs from many companies may usually be had for the price of a stamp and envelope. Updated catalogs will be sent automatically. Talk with your local utility managers and with the automobile dealers. Let everyone who sells or distributes some of the items you need know that you are in the market for certain pieces of used or damaged equipment. Once you have obtained an item, strip it immediately of all usable parts. Store them in appropriate places where they can be cataloged, and then get rid of the remaining pieces which are of no use to you. This will avoid filling your storage area with scrap which will take up room needed for useful components.

Small electronic parts drawers can be obtained and labeled with the parts stored in each. A garage or basement can be shelved to hold the larger components, such as motors, hydraulic pumps, activators, and other mechanical devices. Don't throw anything away which may have some future use, no matter how remote, but, don't keep useless junk around either. This will only serve to confuse the storage arrangement, and sorely needed parts may not be found until months after their use has already been filled by a costly, retail store component.

By practicing resourcefulness, many home applications of remote control systems can be carried out without the expenditure of large sums of money. This book is tailored to the homebuilder and not to the large corporations who can afford to spend millions of dollars on elaborate control systems. Wait, and you may be able to own these multimillion dollar robots for next to nothing when *they* become obsolete and wind up on the surplus market.

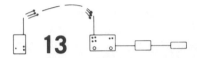

Commercial Remote
Control Systems And Devices

Remote control has become a very popular thing among American families, and systems which provide many of the conveniences of remote control usage are available at reasonable prices to the public. These range from simple tone-activated switches, often marketed as television commercial killers, to highly sophisticated electronic circuits for dialing and answering your telephone from a remote point.

This latter device is shown in Fig. 13-1. This is a Radio Shack model which lists in the two hundred dollar price range. It is a remote cordless telephone system which will enable its owner to answer his telephone or dial numbers and talk through his phone from a distance of up to about 300 feet away. The base sending and receiving unit is attached to the phone line through a mating adaptor and operates from the AC line. The remote unit resembles a Trimline Telephone, is battery-operated, and completely portable. While this remote unit resembles a hand-held transceiver or walkie-talkie, it is a separate transmitting and receiving unit contained in a single, plastic case. From his remote point, the operator can talk to and listen to the party on the other end simultaneously. No push-to-talk switch is used. When the phone rings, a tone is transmitted to the remote circuit by the base station. A switch is thrown on the remote unit, manually, and the conversation is carried on in the usual manner of a hard-line telephone conversation. As long as the remote unit is within range

Fig. 13-1. Radio Shack Cordless Telephone System showing the base unit with the handheld unit at A. This latter device receives charging current from the base. Remote unit alone is at B.

of the base, the caller on the other end will never know that he's listening to a remote control system. At the edge of the transmitting range, some static may be heard, but closer in, no distinction between the remote system and the standard one can usually be made.

Figure 13-2 shows a block diagram of the system operation. Two, highly-separated frequencies are used for transmitting and receiving. The base station transmits on a frequency which lies just above the AM broadcast band in the area of 1.7 MHz. The remote unit receives the base signal at this frequency but transmits back to the base on a frequency around 48 MHz. The base station receives at this frequency and the receiver output is fed into the attached phone line. Again, the base unit transmits on a low frequency and receives on a much higher one, while the remote circuits transmit on a high frequency and receive on the low one. If the transmit/ receive signals were more closely spaced, radio frequency feedback might occur and the receiver of each unit overloaded.

Remote dialing is another matter completely. Figure 13-3 shows how this is accomplished. The receiver at the base has a touchtone decoder circuit. When the touchtone pad at the remote unit is used by depressing the various momentary switches, tones are transmitted and received by the base station which filters the tone and uses the energy in each circuit to trigger a solid-state switch. This action activates an internal dialing mechanism which dials the desired number as commanded by the signal from the touchtone pad. At the remote point, the user hears the click of the dialing and, finally, the ringing of the telephone on the other end. Since the dialing procedure takes much longer than pushing the tone buttons, the base unit stores the numbers commanded through mechanical switches in its cirucit. The numbers are then dialed in the order they were pushed on the touchtone pad.

When the party on the other end of the phone line hangs up, the remote control unit can send a command through a small switch which produces a transmitted signal with the command information for the base unit to disengage the line. When this is done, the base automatically "arms" its circuit, waiting for another phone call, either from the remote station or from a calling party.

Some very complex functions are carried out by this relatively inexpensive system, but each function is very basic and involves much of the theory and practical information presented in this text. Incidentally, circuits such as these are specially licensed by the manufacturer and can be used with your telephone system after

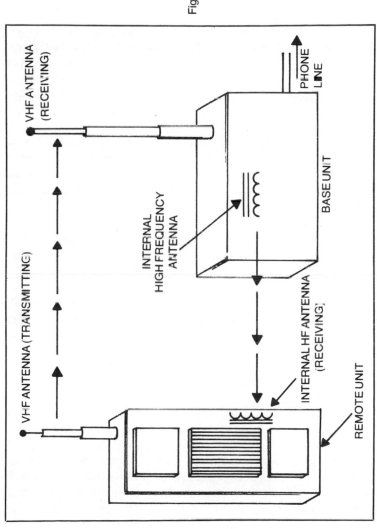

Fig. 13-2. Basic systems operation.

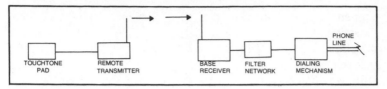

Fig. 13-3. Block diagram of systems operation of Radio Shack cordless telephone.

notifying the utility of its intended use. They will ask you for the unit number and, possibly, for some other information which is contained on the placard which is attached to the back of the unit. Some equipment of this nature is not applicable to all phone systems. This is the reason for notification. With most phone systems, the mating plugs will not require any additional wiring or jacks, but if they are necessary, notify the phone company, and they will perform the needed service. It is against phone company rules to do this yourself.

Figure 13-4 shows a much simpler cordless telephone which can answer but not initiate phone calls. Selling in the $80 price range, the Muraphone provides many conveniences and with a simple, inexpensive circuit design. Putting this unit in remote control terms, it offers fewer channels, and, thus, fewer functions than the previous unit discussed. Operation is basically the same, but this device is a transceiver, using only one operating frequency between the remote unit and the base, and vice versa.

Transmitting and receiving in the 48MHz range, the remote unit has a range of about 300 feet as long as there are no large, metallic objects between the two antennas. When a call comes in, a varying tone, almost like an emergency siren is heard by the person at the remote control point. Keying the remote transmitter sends a tone back to the base unit as before, but this is where this circuit differs greatly from the previous one. A tranceiver cannot transmit and receive simultaneously, so if the base unit is transmitting a call signal to the remote unit, how does the former receive the answering signal from the latter?

The circuit at the base station is designed to transmit at a pulsed rate. This means that the unit transmits for a fraction of a second and then, automatically switches over to receive for another fraction of a second. During a one second interval, the base circuit has switched between transmit and receive many times. It will continue to do this until the signal from the remote unit is detected during one of the brief receive periods. When this

310

happens, a solid-state switch activates a relay which switches the incoming call to the base transmitter input. The caller's voice can then be heard when the remote unit is unkeyed. This is a push to talk-release to listen system at the remote end. The base unit is still switching back and forth between transmit and receive until the transmitter at the remote point is once again keyed. At this point the base goes into full receive until the talk button is released at the remote point.

The result of this system is that the person at the remote location hears the caller's voice in a pulsed manner. The transmitting pulses are so quick between on and off or transmit and receive that the received audio at the remote unit seems to have only a slight waver, but the caller should notice no difference between this and a normal, hard-line phone call, because when the remote station transmits, the base stays in full receive.

This is a simple system, and the audio quality as heard by the caller will not be quite as good as with the previous system. It will sound like the remote operator is talking on a walkie-talkie, because this is exactly what he's doing. The low price of this system makes it a worthwhile device for the person who does not wish to initiate calls and who is not concerned with perfect audio quality. If this is of concern, you'll have to pay more for it by purchasing a more expensive unit.

One note of caution, as was mentioned previously; all of these cordless telephones are licensed through the FCC by their

Fig. 13-4. Simple cordless telephone system for receiving calls only.

manufacturers. While it is possible to build a similar system at home, it is not legal to connect it to your phone line without having prior approval from the phone company and, possibly, the Federal Communications Commission. Most phone companies frown heavily on homebuilt devices being connected to their lines unless they install the proper decoupling isolaters.

One interesting use for this latter cordless telephone would be the controlling of various functions by means of the telephone line and, specifically, through control applications by means of counting the number of times the phone rings. For example, the Muraphone engages a relay when the phone rings the first time. It is conceivable that this component could be replaced with a stepping relay which would advance with each ring. An earlier chapter dealt heavily with stepping relays and suggested methods for remote control using these devices to trigger solid state switches and power relays.

With the circuit modification to the cordless telephone and the stepping circuit from the earlier chapter, it might be possible to control certain functions from the other side of the world by dialing your phone number. Two rings could activate an outdoor light, for instance or even start your video-tape recorder to catch a desired television program. If the system were complex enough, it could even drive the video tuner to select the desired channel. This is mentioned for speculation purposes only and is purely an academic discussion. While all of the rules and regulations have not been checked, it seems obvious that the modification to the circuit would nullify the manufacturer's licensing of the unit. There would probably be problems with the phone company, too, even if you could get their approval on your modified circuit. They're in business to make money, especially on long distance calls. Using this theoretical circuit, you could accomplish a lot of work by ringing a phone which no one answered. Thus, there's no charge to the calling party. Besides, someone might accidentally dial your number, assuming it's a private line with an unlisted number, and throw every work load connected to the system into operation.

This last discussion was intended to stimulate the reader's thinking after having been introduced to a type of remote control system of standard design, the cordless telephone. If you are constantly thinking of modifications, improvements, and whole new methods of operating existing circuits, you will be able to see the potential in a whole new area for devices, originally designed to fulfill a completely different function.

REPEATERS

Another form of remote control through radio signals can be found in the communications repeaters. These are transmitting and receiving stations, combined, which are activated by a radio signal on a certain frequency. Upon receiving this signal at the receiver, the audio output is fed into the repeater transmitter, and the original signal is rebroadcast on a different frequency. Figure 13-5 shows a block diagram of a repeater which works in very much the same fashion as the first cordless telephone discussed in this chapter. A signal is broadcast from a low-lying area to the repeater receiving antenna located at a high point. Usually these transmissions are for communications purposes and the signal will be modulated with the operator's voice. When the transmitted signal is detected by the receiver, its squelch is broken which automati cally keys the repeater transmitter. From this point on, the receiver output feeds the transmitter input. A low-power signal transmitted from the remote side has suddenly become a powerful signal, broadcast from atop a mountain.

The remote radio operator has his transmitter tuned to the input frequency of the repeater while his receiver is tuned to the output frequency of the repeater. If he has a separate transmitter and receiver, rather than a transceiver, he can hear his own voice being transmitted at the instant he is speaking. If someone elsewhere answers his call, they will transmit on the repeater input frequency which will result in a transmission from the repeater transmitter on its output frequency. This latter frequency is tuned by the original caller's receiver.

Repeaters are very useful in providing communications of significant power output and from good antenna sites from a remote

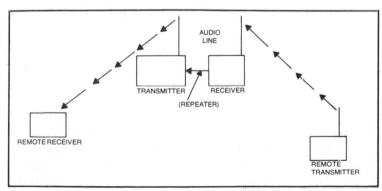

Fig. 13-5. Block diagram of principles of repeater operation.

point, using a low-power transmitter. When mountains interfere with radio signals, a repeater on that mountain or any other close-by high location will enable radio operators to talk with each other through the repeater when they cannot do so by direct transmission methods.

Many repeater stations are run by amateur radio operators for amateur communications. These are most often designed to transmit and receive on the VHF frequencies of 144 to 148 MHz. While providing the basic communication advantages outlined previously, they may even go further and offer many other services. Some provide weather bulletins by retransmitting the audio transmitted by the National Weather Service. Here, a receiver designed to detect these transmissions is set up to have its audio switched to the repeater transmitter on command. This command is sent from the remote point, usually in the form of an audio tone on the transmitter carrier. When this tone is received at the repeater, it supplies energy through a filter network to trigger a switch. This switch connects the audio output of the weather receiver to the repeater transmitter input. The weather information is broadcast and is heard by the station initiating the tone command. When the information has been received, another tone command is sent to return the repeater to its static, non-transmitting state. If this is not done within a set period of time, the repeater automatically returns to this latter mode of operation.

Another repeater convenience offered by many stations is *autocall*. This means that the remote operator can initiate phone calls from his automobile or other remote location. Instead of a weather receiver, a phone line is installed at the repeater site and switched to the transmitter output upon receiving the proper tone commands. When the operator "punches-up" the proper tone sequence during his remote transmission, on switching to receive, he will hear a standard dial tone in his receiver. All of these command tones are usually sent by means of a touchtone pad located at the remote transmitting point. When the dial tone is received, the transmitter is once again keyed, and the correct sequence of numbers corresponding to the phone number being called is punched in on the pad. In receive again, the operator will hear the ringing of the called number being transmitted by the repeater. When the party answers, a normal or fairly normal conversation can be carried on using push-to-talk methods at the remote location. Figure 13-6 shows a Heath transceiver which is especially designed for repeater operations and for autocall. The Touchtone pad is built into the microphone. When the operator

314

desires to activate a service or dial a phone number, he simply presses the correct button on the pad just like he would do on his home phone system. The audio tones produced by the pad are transmitted by modulating the transceiver's carrier output. The output power of this transceiver is only ten watts, but this is more than adequate to reach most repeaters within useful range. His transmissions at low power from a remote point are re-transmitted at a hundred watts or more and, usually, from an ideal location on some mountain top. Repeaters may also offer many other services not fully discussed here. They include emergency police and rescue squad connections, traveler's bulletins, and location services and information.

In many ways, the use of the amateur repeater for autocall parallels the use of the cordless telephones. It should be understood that amateur operators are not permitted to use their equipment for anything other than emergency and pleasure communications. For instance, the autocall feature of the repeater under discussion could not be used to transact business by mobile radio. The telephone companies provide this service through their commercial units which carry a healthy fee. Business band repeaters are operated by private enterprises for business calls. These require that the operators be individually licensed by the

Fig. 13-6. Heathkit HW-2036A 2-meter transceiver showing Touchtone ™ pad on microphone.

FCC and that the transmissions be for business purposes only. These radios cannot be legally used for pleasure communications.

COMMERCIAL RADIO STATIONS

Remote control and automation are crucial aspects of many commercial radio stations. While remote control is often used within the confines of the radio station to remotely activate and control various audio and monitoring devices, one of the most obvious examples involves situations where the radio transmitters and antennas are separated from their audio feeding studios by considerable distances.

Figure 13-7 shows a block diagram of how audio is sent from the studio to the transmitter site and how remote monitoring of transmitter readings is accomplished back at the studio. Audio from a turntable or microphone at the studio is fed to the input of a microwave transmitter. This signal travels on the microwave carrier to a receiving antenna at the transmitter site. Here the audio is recovered by the receiver whose audio output is fed to the input of the broadcast transmitter.

The FCC requires radio stations to monitor and log the various readings of the transmitter on a regular basis. This must be done many times each day. In order to accomplish this at the studio without having to travel to the distant transmitter site, *telemetry* is used. Telemetry is simply a term which means metering from a distance. The transmitter parameters are changed into varying audio tones and fed to the input of another microwave transmitter at the transmitter remote site. These signals are superimposed on the carrier and broadcast back to the studio. The receiver demodulates the microwave signal to recover the audio which is fed into a metering panel. The complex circuitry here converts the transmitted signal into driving current for the various meter indicators. Figure 13-8 shows one of these panels used at an FM radio station.

Microwave transmitters are often used for this relay and telemetry work, but standard telephone lines can also be used to take their place. The principle is still the same but with the audio being transmitted and received by telephone lines instead of microwave transmitters and receivers.

Remote control and automation can operate a radio station without anyone actually being present. Many FM radio stations which are also part of an AM station use automation and remote control to forego the hiring of a separate staff. Most of the staff

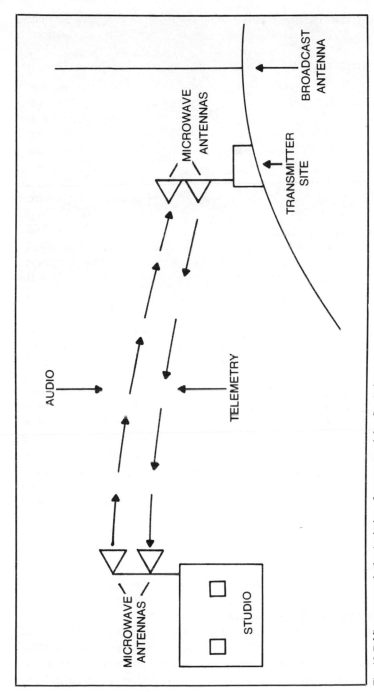

Fig. 13-7. Microwave relaying technique for commercial radio stations.

317

Fig. 13-8. Metering panel from radio station transmitter/receiver link.

works mainly with the AM operation while automation equipment controls the FM. An operator is still legally required to monitor the automated operation.

Figure 13-9 shows a portion of the main automation station with the operator changing a tape reel. Most of the music is pre-recorded and three tapes may be pre-programmed for rotation of play on the air. The master system is programmed from a central control point which is a remote control unit using a hard wiring system for connection. Here the operator is programming certain functions into the master unit from his console. A continuous clock allows certain functions to occur at a specific time. This console offers much versatility and is relatively simple in design considering the many functions it must perform. Figure 13-10 shows a close-up of the console with some of the function switches indicated.

Radio commercials, weather reports, and other information which must be continually updated are usually recorded on audio cassettes which are installed in coded slots in a motorized portion of the main automation system. When a certain cassette is called for at a pre-set time by the programming, the motorized cartridge chain begins to rotate, bringing the correct cassette into a tape player on command. Figure 13-11 shows the tape cartridge panels.

This is a very complex system, but it is made up of many of the circuits already discussed but in much greater abundance. Figure 13-12 shows a small portion of the internal circuitry in the main control system. All of the circuit parts are installed on removable, printed circuit boards. If a malfunction should occur in one of the components, the non-functioning circuit can be traced to the appropriate card or printed circuit board. The board is then replaced with a spare until repairs can be made on the original. This is a must for commerical radio stations who cannot afford to be off the air for long, because what they sell is radio time. A fairly

Fig. 13-9. Operator changing tape reel on main console of commercial radio station automation equipment.

Fig. 13-10. Closeup of automation console control unit.

complex power supply is also housed in the main system with many different circuits and feeds coming off of it. Figure 13-13 shows a portion of the supply which must provide many different operating voltages at various current levels.

One other FCC requirement applied to commercial radio stations is that of logging all commercials, sports events, remote broadcasts, etc. This automation system has an automatic logger which prints out in hard copy the programming events of the day. Its input is in the form of coded signals received from the main control system. The codes are compared with their corresponding meanings which have been internally stored through another programming procedure at a second input to the printer. The resulting printout indicates what has occurred, when, and how many times.

This system provides many operations which might require several individuals to accomplish without automation. It takes the place of the announcer, engineer, and program logger, but not completely. Anything mechanical or electronic can malfunction. When this occurs or a programmed event has been missed, the system sounds an alarm which indicates a situation has occurred which is beyond its control. Here is where the human element must come and bail the machine out of its difficulties. This last statement may not really be fair to the machine because the missed event probably occurred due to a human programming error.

320

COMMERCIAL CARRIER TRANSMITTERS AND RECEIVERS

Power utilities require extremely close monitoring of their electric lines to indicate line conditions in their highly complex systems. This monitoring is usually accomplished through carrier current transmitters and receivers which relay signals and commands over the electric lines. The problems encountered are far

Fig. 13-11. Rotating tape cartridge panel of automation console.

Fig. 13-12. Circuitry of small portion of automation console.

more severe than those discussed in the chapter of this book on carrier-current transmitters. This is due to the highly complex transmission network which takes in hundreds and even thousands of miles of electrical conductors. Another major problem is in coupling to the line which carries tens of thousands, or even

Fig. 13-13. Central power supply that provides operating voltage and current to the entire automation system.

hundreds of thousands of volts. Capacitors which could be mounted in small aluminum boxes in the earlier chapter of this book now take on mammoth proportions to block against the high potentials presented by these lines. Figure 13-14 shows one of these capacitors.

One major use of the carrier-current signals is in providing line disconnect relays with signals to activate during a fault or grounding of one of the lines. When this occurs, a tremendous

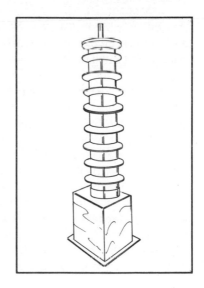

Fig. 13-14. Extremely high-voltage line isolation capacitor for carrier-current control.

amount of current will flow which may be measured in thousands of amperes. This surge will cause a back-up circuit to respond should the carrier-current control fail, but this latter even takes time (measured in fractions of a second) which can cause insulators to heat and explode. Millions of dollars worth of line components can be quickly destroyed by this immense current drain.

Figure 13-15 shows a typical electric line circuit which uses carrier-current transmitters and receivers to detect line faults and shut down the entire section of the line affected while routing power through alternate lines. This enables the company to supply power to areas which might, otherwise, be without power. The only customers affected will be those on the section of line which has the fault. This fault, however, will not conduct power from other portions of the overall system if the carrier current controls work properly. The various elements in this complex connection system are indicated in the drawing.

Figure 13-16 shows the applications of transfer trip relaying. This is a double back-up system which removes a separate line coming off of the one which has incurred the ground fault. When the grounding is sensed by the line relay nearest the fault, transmitter T2, in this instance, is activated, sending a signal to receiver R2 to disconnect relay RL3 which connects the second line, line B to line A through a transformer. When RL3 disengages the second line, it is not subject to damage from the ground fault at line A. The fault will probably be sensed first at RL2 because the fault lies closest to this component. RL1 will also sense the fault a split second later. This will activate transmitter T1 which will send a signal through the line to receiver R1 which also commands RL3 to open its line contacts. If T2 and R2 should fail to operate properly, T1 and R1 act as a back up in this situation.

This has been a very simple explanation of a highly complex matter. In actual practice, line B will automatically be tied in to another feeder line by signals sent down other lines from the carrier system. When the fault is sensed, commands may be sent hundreds or thousands of miles to tell other systems to shut down and still others to reroute their power connections.

Expanding the system further, we see in Fig. 13-17 that another carrier transmitter is located at the transformer. The transformer feeds from line A. Two other receivers are located at the ends of line A. Line B is connected as before. Should a fault occur within the transformer, its transmitter picks up a signal from the transformer relay and transmits a command signal to the two

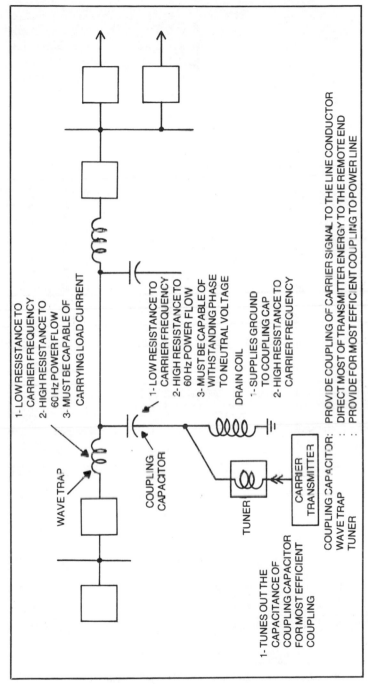

1- LOW RESISTANCE TO
   CARRIER FREQUENCY
2- HIGH RESISTANCE TO
   60 Hz POWER FLOW
3- MUST BE CAPABLE OF
   CARRYING LOAD CURRENT

1- LOW RESISTANCE TO
   CARRIER FREQUENCY
2- HIGH RESISTANCE TO
   60 Hz POWER FLOW
3- MUST BE CAPABLE OF
   WITHSTANDING PHASE
   TO NEUTRAL VOLTAGE

DRAIN COIL

1- SUPPLIES GROUND
   TO COUPLING CAP
2- HIGH RESISTANCE TO
   CARRIER FREQUENCY

WAVE TRAP

COUPLING
CAPACITOR

TUNER

CARRIER
TRANSMITTER

1- TUNES OUT THE
   CAPACITANCE OF
   COUPLING CAPACITOR
   FOR MOST EFFICIENT
   COUPLING

COUPLING CAPACITOR: PROVIDE COUPLING OF CARRIER SIGNAL TO THE LINE CONDUCTOR
WAVE TRAP : DIRECT MOST OF TRANSMITTER ENERGY TO THE REMOTE END
TUNER : PROVIDE FOR MOST EFFICIENT COUPLING TO POWER LINE

Fig. 13-15. Operation of wave trap, coupling capacitor and tuner in commercial power carrier-current system.

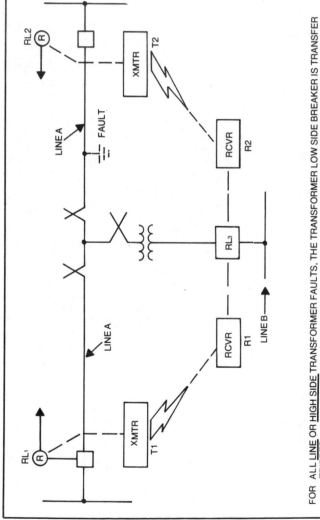

Fig. 13-16. Applications of transfer trip relaying.

FOR ALL <u>LINE</u> OR HIGH <u>SIDE</u> TRANSFORMER FAULTS, THE TRANSFORMER LOW SIDE BREAKER IS TRANSFER <u>TRIPPED</u>
FROM EITHER OR BOTH REMOTE LINE TERMINALS

PURPOSE: TO INSURE THAT ALL BREAKERS PROVIDE HIGH SPEED CLEARING OF FAULT IN ORDER TO ACHIEVE
SUCCESSFUL HIGH SPEED RECLOSING ON THE LINE TERMINALS

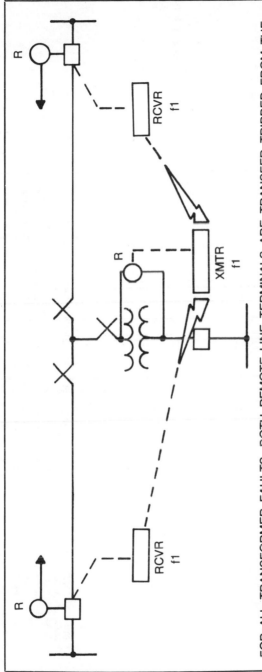

FOR ALL TRANSFORMER FAULTS, BOTH REMOTE LINE TERMINALS ARE TRANSFER TRIPPED FROM THE TRANSFORMER LOCATION

PURPOSE: TO INSURE THAT THE FAULT CURRENT SOURCES TO THE TRANSFORMER ARE TRIPPED OPEN BECAUSE THE LINE TERMINAL RELAYS MAY OR MAY NOT OPERATE FOR THE VARIOUS TYPES OF TRANSFORMER FAULTS

Fig. 13-17. Transformer fault activation through carrier-current control.

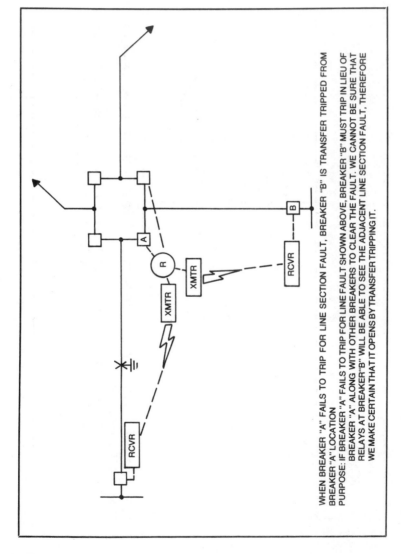

Fig. 13-18. Breaker failure relaying operation.

WHEN BREAKER "A" FAILS TO TRIP FOR LINE SECTION FAULT, BREAKER "B" IS TRANSFER TRIPPED FROM BREAKER "A" LOCATION

PURPOSE: IF BREAKER "A" FAILS TO TRIP FOR LINE FAULT SHOWN ABOVE, BREAKER "B" MUST TRIP IN LIEU OF BREAKER "A" ALONG WITH OTHER BREAKERS TO CLEAR THE FAULT. WE CANNOT BE SURE THAT RELAYS AT BREAKER "B" WILL BE ABLE TO SEE THE ADJACENT LINE SECTION FAULT, THEREFORE WE MAKE CERTAIN THAT IT OPENS BY TRANSFER TRIPPING IT.

receivers to transfer trip that section of the line. The same occurs to line B which is also transfer tripped. This removes both lines from the fault until repairs can be made and the relays re-set. At this point, the transmitters and receivers are returned to the armed positions awaiting another fault to occur. Figure 13-18 explains another use of the carrier current system in commercial power applications. This is called break-failure relaying and depends upon the carrier system to act when a circuit breaker fails to do so.

SUMMARY

Remote control has been with us since the dawn of man. When the cave man used a tree branch to wedge under a large stone in order to move it, he invented a simple machine which was able to transfer his human strength to the work load and to earn himself an advantage through force multiplication. Today, remote control is to be seen everywhere. Department stores, residences, factories, and many other businesses and concerns take remote control for granted, because its use is so common.

Without the commercial uses of remote control, many of the services which are rendered on a daily basis and considered crucial for life as we know it would be utterly impossible. This chapter has barely touched on the wide range of remote control used in business and industry. Complicated as these systems may be, they all have their roots in the basic ideas of control presented in the pages of this book. The commercial systems are but enlargements and slight modifications of many of the devices and principles already discussed. Taken on a section by section basis, these large, multimillion dollar systems are all comprised of simple building blocks of remote control circuits and machines.

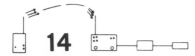

14

Remote Control Applications

As a culmination of all of the materials presented in this book, it is time to begin the discussion on direct application of remote control for practical purposes in and around the home. Perhaps you have already thought of or discovered many uses for the circuits, machines, and theories already presented in this text. This chapter will provide you with some more ways in which various work functions can be controlled remotely. Again, the purpose of this book is to provide the experimenter with information on the usual assortment of tools and equipment to learn of and build remote activation equipment. For this reason, all of the projects have been relatively inexpensive and practical for many persons. The concentration could have been on the elaborate remoting systems used in and around large factories, but a discussion along these lines would have provided the reader with a lot of theory and very little he could build in the home shop.

Most of the systems in this chapter have been built and operated with success in and around the home. Many of these suggestions will be directly applicable to your situation, some may require slight modifications, others, a great deal of modification, and still others will not be applicable at all. The systems which do not fit your needs may, however, give you some ideas for developing an alternate system along the same lines but for a different purpose.

GARAGE DOOR OPENER

Figure 14-1 shows a block diagram of a system for the control of garage door opening and closing. The system uses one of the

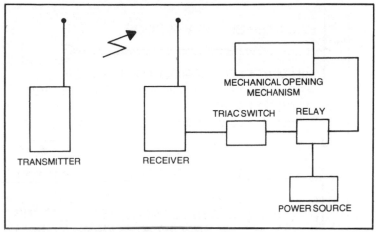

Fig. 14-1. Block diagram of radio-controlled garage door opener.

many commercial openers on the market but saves money by building the remote control circuitry in the home shop. What you should purchase to go this route is an electric garage door opener which is designed to operate from 120 volts AC and be switched on or off by a switch inside or outside of the garage. You will find that these systems are far less expensive than the radio controlled garage door openers which contain the same mechanical components and a few simple electronic circuits. *You* will be providing the circuits in this project.

Figure 14-2 shows a top and side view of an installation detail for a sectional door. In this type of installation, the motor assembly and chain-drive track attaches to ceiling or wood support with angle straps. It should be located in the exact center of your garage door opening with one end of the track attached to the header above the garage door opening using the mounting bracket that comes with each kit. The door-connecting arm that is shipped assembled to the track rail is adjustable for door height clearance. Merely loosen the two bolts, adjust the mounting bracket section to the proper position and secure it to the top of your garage door with the two or three wood screws that come with the outfit. Run a 120-volt power line from your panel box, or nearby lighting or convenience outlet.

A typical installation would begin by ordering a door system designed for your particular type of door, that is, upward-acting single or double sectional doors or a one-piece door with track or jamb hardware. A sectional door installation is shown in Fig. 14-2, a one-piece door installation is shown in Fig. 14-3, and a one-piece

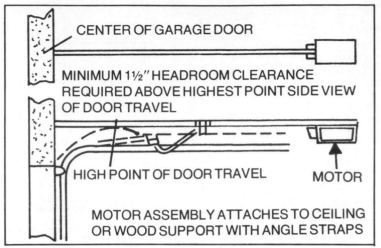

CENTER OF GARAGE DOOR

MINIMUM 1½" HEADROOM CLEARANCE
REQUIRED ABOVE HIGHEST POINT SIDE VIEW
OF DOOR TRAVEL

HIGH POINT OF DOOR TRAVEL   MOTOR

MOTOR ASSEMBLY ATTACHES TO CEILING
OR WOOD SUPPORT WITH ANGLE STRAPS

Fig. 14-2. Top view and side view of a garage door opener installed on a sectional door.

door with jamb hardware is shown in Fig. 14-4. In all situations, make certain that you have enough head room for the track before ordering the system.

Since the one-piece door with jamb hardware is the most complicated to install, we will use this type as our installation example. Before beginning any mounting of the automatic door system, operate the garage manually several times in order to determine the lowest and highest points of the door top as shown by the dotted line in Fig. 14-4. When you are certain that you have correctly determined these points clearly mark each one.

Next line up the motor housing in the exact center of the garage door and level with the lowest point of the open door. Position it only, do not secure firmly. Elevate the track or T-rail end that is opposite the motor housing to an angle required to clear the highest point of the door's path of travel. Using the screw holes in the mounting bracket as a guide mark the location of the screw holes on either the ceiling or wall, whichever the bracket is to be mounted to. In choosing this location make certain that the structure is strong enough to mount the bracket.

If the structure to which the bracket is attached is wood, conventional wood screws supplied with the kit will suffice. However, if the attaching point is masonry, toggle bolts will have to be used for mounting to hollow concrete block, and lag or other type of masonry anchor will have to be used in brick or solid masonry walls or ceilings. For the former (toggle bolts) drill a hole through the masonry to the hollow area, large enough for the bolt and wings to pass. In solid masonry, drill a hole with either a masonry drill or star drill and hammer. The size of the hole should be just large enough to accept the lead sleeve and deep enough so that it fits flush with the wall or ceiling surface prior to setting the anchor. A special tool for setting the anchors is sometimes supplied with them. The method of installation is shown in Fig. 14-5.

Once the door end of the track or T-rail is secure, mount the motor housing in its proper position (level with the lowest point of the door top as it travels). Where the ceiling consists of wood joists the motor housing can be secured directly to the sides of the joists with wood screws. In masonry walls or ceilings a special bracket will have to be secured to the wall or ceiling first with lead anchors or toggle bolts before the motor housing can be attached to it with bolts and nuts.

The next logical step would be to mount the connecting arm from the track of the T-bar to the door. Mount this arm directly in the center of the garage door and near the top of it. Wood screws are provided with the kit for installing the arm to the garage door, and bolts and nuts are used to connect the arm to the track of the T-bar. For a one-piece door with jamb hardware, a longer-than-usual arm, about 4 feet long, designed to operate this type of door more smoothly and efficiently should be used.

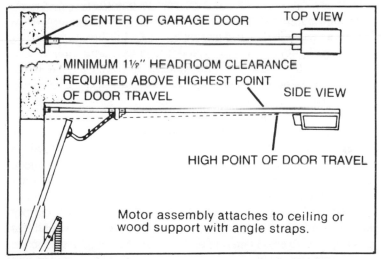

Fig. 14-3. Top view and side view of a garage door opener installed on a one-piece door.

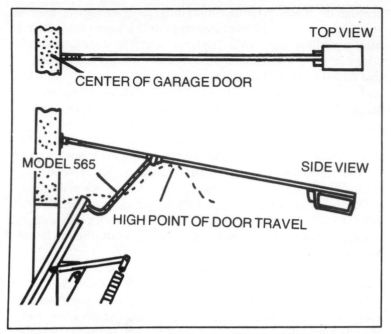

Fig. 14-4. Top view and side view of garage door opener installed on one-piece door with jamb hardware.

Drive motors used on most residential garage door systems usually are rated between one-quarter to one-third horsepower, each pulling approximately 5 amperes of current. Therefore, AWG No. 14 will handle this load quite well. In fact if the circuit feeding the garage lights is not fully loaded it is quite permissible to feed the drive motor of the operating system directly from one of the nearby lighting outlet boxes.

Of course, if you already have a motorized garage door opener installed at your home, all of the preceding instructions are done away with, and you will simply add the remote control circuits to control what you already have. The installation of the door and the mechanical equipment will be the most difficult part of this project. The electronic control is very simple and can be handled in any number of ways.

The author's garage door system worked on AC house current and required manual closing, so a single channel unit was all that was required to effect one remote function, the switching of power to the motor for opening. Figure 14-6 shows that radio control was elected for the remote system, and two, inexpensive walkie-

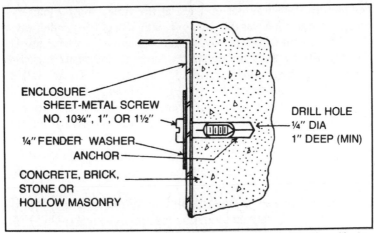

Fig. 14-5. Method for securing parts to masonry walls. First, drill hole with star drill or masonry bit. Insert the lead or plastic anchor. This should be snug in order to hold properly. Finally, with the attachment in place, insert the sheet metal screw or lag bolt.

talkies operating in the 48-MHz frequency range were brought into play.

The receiver operated from a nine volt battery which was replaced with a rechargable type and kept in a constant state of charge with an inexpensive charger unit sold with the battery. Alternately, a separate, nine volt supply could be built and operated from the 120-volt line, doing away with the battery

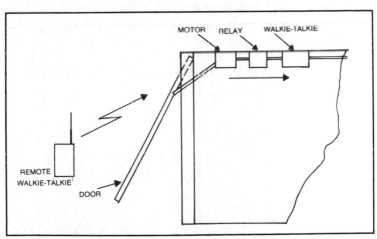

Fig. 14-6. Block diagram of control system showing transmitter and receiver mounted to ceiling.

completely. One was not on hand at the time, and the battery-charge combination was the quickest method of getting under way.

The output of the receiver was connected to a triac switch directly with no tone filter due to the remote location of the garage from surrounding homes and other radio traffic. These walkie-talkies are low power and have a very short range. The antenna of the receiver was completely lowered, so receiver sensitivity was brought to an extremely low level. It was necessary for the transmitter to be very close to the garage door to effect the triggering. This situation was brought on purposely to avoid the possibility of other transmission on this frequency from causing false, unwanted openings.

Figure 14-7 shows the switching circuit which is fed directly from the receiver with no filters in between. A triac was chosen to operate a 120-volt AC relay directly rather than having to resort to a small DC relay and, possibly, a separate DC power supply. When triggered, the switch feeds power to the relay which, in turn, connects the power leads to the garage door motor. When the door has fully opened, the switching function is stopped by stopping transmitter operation. This particular system had an automatic cutout on the motor control which automatically stops all power to the motor when the door had been opened to a certain point.

Operation

Once this system is set up and working, all that is necessary to open the garage door is to pull the automobile up to the garage as closely as possible and trigger the push to talk button on the transmitter. The receiver used was extremely noisy when the squelch was broken by the transmitted signal, so it was not necessary to provide any audio input to the transmitter. Less noisy receivers may require you to hum into the transmitter microphone

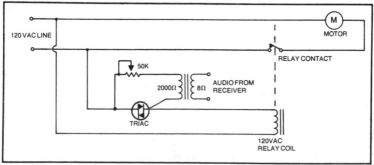

Fig. 14-7. Schematic diagram of triac control circuitry.

for the few seconds it takes to open the door. Some transmitters have a built-in tone generator for sending code. By locking the code key or button into the sending position, a steady tone will be broadcast which should easily trigger the circuit. Sensitivity of the triac triggering circuit can be adjusted with the variable resistor.

Alternately, a latching relay may be used to trigger the 120-volt line to the motor and only a short burst of transmitter power will be needed to set the system in motion, providing that the motor cut-off switch can be made to reset the relay. Here, sound activation might be used with the sound being produced by the car horn. The transmitter may be placed in a weather-proof box mounted to the garage door frame as shown in Fig. 14-8. Since a transmit volume control is rarely included in simple walkie-talkies, it may be necessary to fabricate a mechanical counterpart by taping padding to the microphone/speaker combination. Another battery charger will have to be attached to the transmitter power supply and this unit left in the transmit position at all times. Then, when you want to open the garage door, all you do is pull up to the front, blow the horn, and the opening process begins. It will be necessary to set the sensitivity control so that the mechanism will operate only when the transmitter input receives a *loud* horn blast. This will be transmitted to the receiver which will act on the demod-ulated audio output to send a control signal to the switching circuit.

ALARM SYSTEMS

Burglar and intruder alarm systems fall under the category of remote control. The intruder is the unwilling controller in these applications and sets off a single or series of remote control functions when he enters a home (if the system is set up properly). Many of these systems can also double as fire and smoke sensors which are brought into play to trigger similar alerting devices.

One of the main problems encountered with alarm systems is the power supply. If powered from the AC line, all the would-be thief has to do to get in and out undetected is to cut the conductors of the service to the home. The system will then be rendered completely ineffective. If twelve volt DC power is used, this may be derived from a battery, but even though the current drain may be very low, the battery could eventually discharge if power were to fail for a long period of time while you were away on vacation.

Figure 14-9 shows a method which can practically guarantee that you will always have power when you need it, because the power supply can come from either the AC line or from a separately

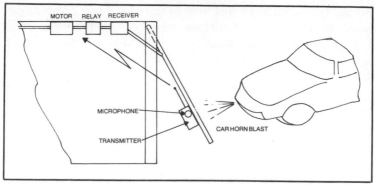

Fig. 14-8. Block diagram of car horn activation through transmitter control.

control in the SCR switch will allow triggering sensitivity to be set here. If an amplifier volume control is desired, the 15K ohm fixed resistor at the base of the transistor may be replaced with a 100K ohm variable resistor.

It will be necessary to set the triggering sensitivity control after the volume control of the amplifier is adjusted if this latter component is used. Trigger the alarm tone at the remote location, and adjust the SCR variable resistor for firing. After, the SCR fires, turn off the remote alarm and depress the button on the switch located in the SCR circuit. This will remove the power from the SCR and allow it to go back to its resting, non-conducting state again. Now, trigger the remote alarm again to make certain the contained storage battery. A battery charger is connected to the AC line and constantly supplies charging current to the battery which powers the 12-volt alarm and the sensing system throughout the home. Now, if the power should fail while you're away, there will be many days of operation left in the fully charged storage battery, assuming the sensing circuit relay does not draw a lot of current. This relay is the normally closed variety and can be a latching type if desired. When the system is armed, the relay is activated and the contacts are held apart. When an intruder breaks one of the sensing contacts, the relay loses power, the contacts close, and power from the charged battery activates the alarm.

This is an excellent system, but it is necessary to hard-wire all of the sensors to the relay circuit. Some persons may have storage sheds and other outbuildings which are located a sizable distance from the home. Here, some form of remote control is desirable to get a command signal back to the central alarm system. Again, transmitter control is the logical means.

A similar system to the garage door opener can be used with walkie-talkies forming the carrier/receiver portions of the circuit, but an even better system might use audio tone control as a safeguard. The system is shown in Fig. 14-10 and uses a small sensor loop system composed of a single, normally open switch contact. This can be attached to the door or window, so when an intruder enters, the sensor automatically closes its contacts. This completes the circuit to a battery powered tone generator and audio amplifier. A loud audible tone is produced and amplified. This should be enough to frighten most intruders off, but if you're away and he knows this, it might not be enough, especially if you live in a remote location with no surrounding homes.

This system goes further. The amplified tone is picked-up by an audio sensor at the home. The audio sensor feeds the received tone through a preamplifier which passes the tone on to a solid-state switch. This, in turn, opens the contacts of a relay which is placed within the normally closed sensor system of the home. The home alarm now sounds and lights may be turned on within the home as well if they are attached to the main system relay in the correct manner.

Figure 14-11 depicts the sensor circuit at the home and shows the various circuits the received tone must travel through to trip the central alarm system. A simple carbon microphone element is used and may be salvaged from an old telephone headset. The transistor acts as an amplifier, bringing the low-level microphone output up to a value that will cause the SCR to fire when it receives the amplified output. The relay does not need to be a latching type, because the SCR will continue to conduct its DC input, upon firing, until power is removed from the system. No volume control was included with this circuit at the amplifier, because the variable

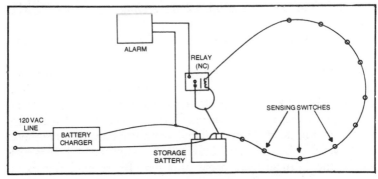

Fig. 14-9. Intruder alarm system showing various components.

SCR will fire from this command without further adjustment. If this is necessary, adjust by a few increments at a time and repeat the process until positive triggering of the SCR and its central alarm circuit is obtained every time.

Other Ideas on Alarm Systems

While not specifically remote-control oriented, the alarm itself can be a very complex device which can perform several, discrete jobs. The author designed and built an alarm network for attachment to the rest of the intruder system which went beyond the capabilities of most alarms today. This rather simple but effective alarm was necessary because of the author being away from home for several weeks at a time. The intrusion system was identical to the one discussed earlier with a 12-volt storage battery for a long-lasting power source should the charger's AC power fail. The sensors, relays, alarms, etc. were all designed to operate from 12 VDC and, thus, were not dependent on the AC line for operation. The difference in the author's system was evident when it was triggered. Instead of the tripping function turning on a light or a siren, it performed a combination of these.

Flashbulbs used, normally, for photography work, were placed in small sockets which were fabricated from aluminum. They were placed at various locations throughout the home. In all, some 25 flashbulbs were used. Additionally, 12 VDC relays were installed on the power cords to the stereo system and to several lights throughout the dwelling. The *coup de grace* was a tape recorder in the basement with a pre-recorded tape of a vicious dog barking and a heavy masculine voice say, "Hey, who's up there?" Several other alarms were also included.

Figure 14-11 shows how the system was set up. When any sensor in the loop opens up, pandemonium breaks loose. The intruder would be bombarded by the flashbulbs firing in all areas of the house, temporarily blinding him. The siren goes off as the overhead lights illuminate. To add to all of this, a telephone bell begins to ring, the stereo (which was pre-set to a high volume) blares, the doorbell starts ringing, and the dog starts barking. All of this added to the recording of the human voice in the basement should serve to drive away the most professional thief. The reason for the multitude of alarms is found when you consider that there are several points of entry into a home. One never knows just where the intruder may break through. By placing alarms in various sections of the home, the intruder is sure to get the

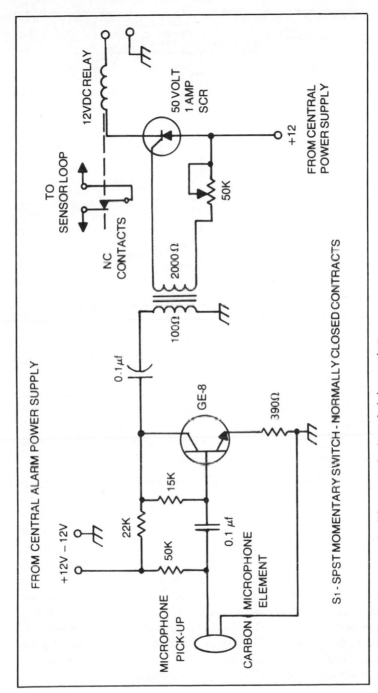

Fig. 14-10. Tone alarm sensor circuit for connection to central alarm system.

341

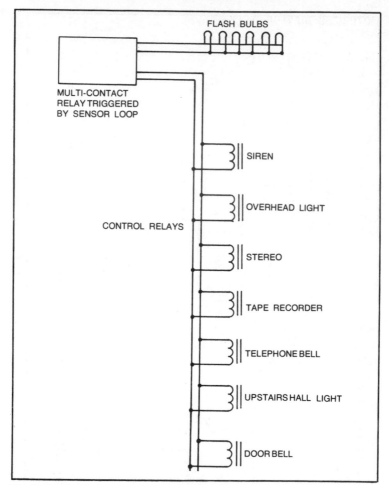

Fig. 14-11. Multifunction alarm system wiring.

message. To him, blinded by the flashbulbs, it must seem that the entire world is having a party in the home he has just entered. He may never hear the voice in the basement, if he entered through an upstairs window, but he will hear and see many of the other alarms.

Admittedly, this may be going a bit overboard, but the remote, isolated location of the author's home made this system necessary. Incidentally, in the year and a half this system was installed in the home, it was only triggered once. The intruder was the author who arrived early in the morning after a grueling six hour flight. Forgetting to de-arm the system before opening the front door, he was royally greeted upon entering.

342

MOTORIZED DRAW DRAPE CONTROL

For special effects, nothing is quite as impressive as being able to open the drapes over a picture window to allow guests an exquisite skyline view from your home. Those persons who are fortunate enough to live in locations where these panoramic views are available may find that next project to their liking.

Simplicity is accentuated here as far as the design and parts are concerned, but the effect is very dramatic and impressive. The amazing part of this system is that nothing need be changed or modified regarding your present drapery installation. All that needs be done is to add a slow turning motor, some control wiring, and a simple switching panel. An old TV antenna rotor and its control box were called into service for this remote control application which is pictured in Fig. 14-12.

If a pulley arrangement is used to draw the drapes manually, the installation is very simple and can be performed in less than two hours. This even includes the routing of the control cable in most instances.

Begin by building a small platform to hold the motor off the wall. This project mounted the motor at the center of the drawstring, but it can be mounted at the bottom and top as well. A small pulley is attached to the motor shaft which turns at about one revolution per minute. Mount the motor on the platform and secure it. The motor case must not be allowed to move at all or it will soon shake itself loose from the mounting and fall to the floor, possibly taking the drapes and pulleys with it. The pulley should be small

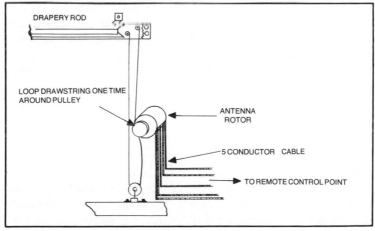

Fig. 14-12. Remote controlled drapery system using antenna rotor for power.

enough to allow the slack in the drawstring to be looped around it once without becoming overly tense. That's all there is to this part of the installation.

This particular rotor had a five conductor cable for control. Others may use a seven conductor system. Some of the conductors are designed to carry signals back to the sending unit to indicate what direction the antenna is pointed toward. If these portions of the cable can be isolated, they can be done away with for these purposes. Route the cable to the desired remote control point. The cable can be run in the attic or even through the conduit designed to carry electrical wiring if this is permitted by local electrical codes. If this latter arrangement is used, make certain there are no nicks in the control cable insulation.

Once the wiring has been brought to the control point, simply connect the original rotor control to the appropriate conductors and wire the unit to the AC power system. The author removed the entire circuit from the bulky case it came in and remounted the works in an attractive, simulated walnut cabinet. The rocker-arm switch was replaced with two lighted, instantaneous switches. Holding one switch in opens the drapes while the other closes them.

Test the arrangement by applying power to the motor with the control unit. The drapes should slowly and evenly begin to open or close, depending on which control you are depressing. If the drapes open too fast, it will be necessary to install a smaller pulley on the motor shaft. If too slow, increase the size of the pulley. Make certain the cord is wound tightly enough on the pulley to prevent undue slippage but not so tightly that the drawstring will not slip when the drapes are fully opened or fully closed. You can judge when to break the power to the motor, if you are watching the operation closely, but if you don't stop in time, the motor can literally pull the drapes and suspension to the floor in a few seconds if the cord will not slip under the pressure presented when the drawing mechanism is at the end of its run.

If problems should be encountered with a too tight coupling between drawstring and pulley, a bit of vaseline on the pulley should take care of the problem, but this may stain the curtains. Alternately, you can increase the length of the drawstring by an inch or two to provide a looser fit at the pulley. Tension adjustments are not usually difficult and the entire system can probably be adjusted in a few minutes.

344

Now, whenever you wish to show off the beautiful view you have from your mountaintop home or penthouse apartment, you can stand at the remote control point, make your speech, and then depress a small switch to verify the grandeur you have been talking about. The motor noise is usually not objectionable and usually is heard as a medium pitched, soft whine.

MOTORIZED HIDDEN BAR

One home the author owned had large bookcases and cabinets in every room. This was much more storage space than needed, so it was decided that one of the bookcase/cabinet combinations would be turned into a hidden, miniature bar. Figure 14-13 shows how the cabinet appeared before modification. Fortunately, this piece of furniture set in a corner and the bookcase was separate from the matching cabinets. It simply sat atop these units, being connected by small clamps. The cabinets had open backs which were enclosed by the wall they were mounted against.

To do what was desired, it was necessary to secure the bookcase to the walls and ceiling. The cabinet which it was resting on was to serve as the hidden bar. While we have been referring to

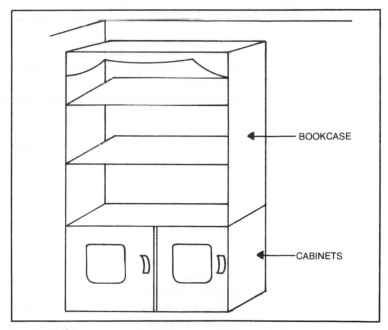

Fig. 14-13. Original bookcase and cabinet arrangement before modifications for hidden bar.

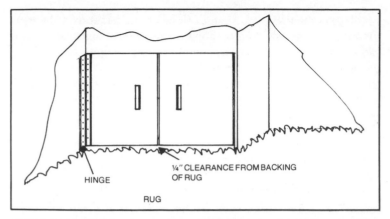

Fig. 14-14. Hinge arrangement of lower cabinet to allow it to swing toward the left wall.

this latter piece of furniture as cabinets, the two units are actually one cabinet with two doors. The bookcase was mounted in the same position by simply hammering nails through the top, back, and sides to mate with the studs in the walls and ceiling. Fancy wood screws might have been better, but a dab of matching brown paint on each nail head camouflaged them nicely.

With the top portion, the bookcase, secured, the cabinet section could be pulled completely away from the wall without affecting the top unit. Figure 14-14 shows that a long, sturdy hinge was run between the wall and the entire height of the cabinet section. A small amount of the bottom surface of the cabinet was planed away which allowed the cabinet to swing freely on its hinge. The bottom edge had about a quarter inch clearance from the floor. This space was hidden by the shag of the rug.

It was now time to decide on the driving mechanism. Every bit of space behind the cabinet doors was needed, so hydraulic rams were out of the question. So were electric motors with complicated pulley arrangements. This had to be a "hidden" bar. No one should ever expect it to be there unless told or shown. The idea finally came after a few hours of intense study. An electric motor would be used but the pulley attachment would be a rubber wheel connected directly to the shaft. The wheel would mount in a cut-out at the bottom of the cabinet and would make contact with the rug. Again, an old antenna rotor was used for the electric motor. Figure 14-15 shows how it was mounted. A rubber children's coaster wagon was partially disassembled for the wheel which was easily mated to the shaft of the motor and secured with the original carter pin. The

346

wiring was run beneath the floor of the room and brought through the wall at the far end where the remote control point was to be. An additional control point was established at the bookcase and hidden on the inside panel.

Testing was surprisingly simple. It was expected that problems would be encountered with positioning of the motor and the drive wheel. In building and installing this section, the motor was simply mated with the wheel and installed on the bottom shelf, mounted to rubber spacers. Several were needed to bring the motor high enough off the bottom shelf to allow the wheel to slip into its cut-out space. The motor was bolted to the shelf through the flexible spacers which were put there to allow for minor wheel tension and adjustments. As the mounting bolts are tightened, the spacers are compressed and the wheel mates more closely with the rug floor. Initially, the motor was set on the spacers so that the wheel touched the floor. The motor was tightened quite a lot, with the author figuring on decreasing tension a bit at a time, but when the circuit was activated, the cabinet swung out with no problems, as smoothly as silk. Various tension adjustments on the wheel and motor showed that this was not critical, because the rubber wheel created quite a bit of friction between it and the surface of the rug. This was the equivalent of putting chains on automobile tires. Everything worked smoothly. Figure 14-16 shows the completed arrangement with the cabinet/bar extended. The bar may be closed again by flipping the small switch inside the bookcase. Make sure the wiring is flexible and long enough where it leaves the cabinet, enters the case and goes beneath the floor. It will be flexed as the cabinet is motored out or in.

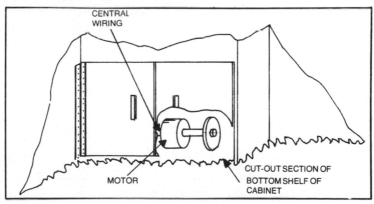

Fig. 14-15. Cutaway view showing motor and power wheel arrangement.

347

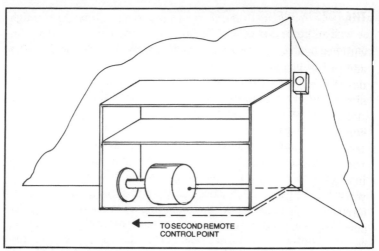

Fig. 14-16. View of finished project showing cabinet in out position.

Antenna rotor motors are excellent sources of rotary power for remote control applications. They are geared for very few rpm and possess high torques. Large pieces of furniture can be moved by even the smaller rotors. They do come in all sizes, however, and can be quite inexpensive if you look around television sales and installation outlets that install and remove TV antennas.

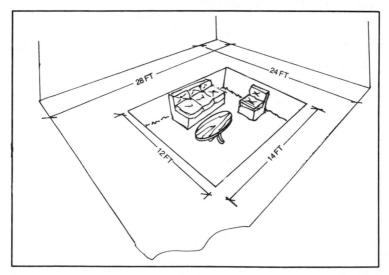

Fig. 14-17. Overall view and hypothetical measurements of sunken floor section.

THE TOPS IN SOPHISTICATION

For those readers who are blessed with very large homes and with money to burn, here is a hypothetical project which, admittedly untested, should work. The author was prompted to design this system after he saw a spy movie which had something similar in it. I guess it would be called an intimate conversation nook. This differs from a sunken living room, but it could be called a *sinkable* living room. The idea is to be sitting cozily in a section of a large, carpeted room, then, on command, cause this portion of the room to begin to sink to a predetermined level. The result is shown in Fig. 14-17. The easiest way to control the size and weight of the work load, 12-foot by 14-foot section of flooring and all of the furniture that goes with it, would be to use a hydraulic lift purchased from some out-of-business service station or garage. A contractor could probably handle the cutting and reinforcing of the floor which would be necessary. The automobile rack could be left in place and the floor bolted to it. The hydraulics would be located in the basement directly beneath the section of flooring to be raised and lowered. This is shown in Fig. 14-18. The control panel would be mounted on the movable floor section, attached by flexible hydraulic lines. Carpet would be used on the entire room floor and movable floor section. This same carpet would extend down all sides of the floor well. The shag of the carpet could probably hide the fact that a movable floor section had been installed by

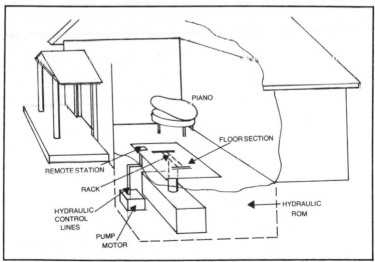

Fig. 14-18. Side view of hydraulic system attachment for sunken floor section.

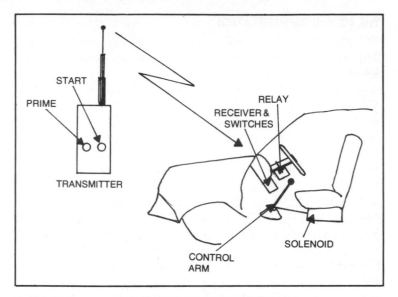

Fig. 14-19. Component placement for remote car starting.

overlapping with the movable section carpet when the latter was raised to a level position with the rest of the floor. If clearances were alright, all kinds of fancy aquariums, stereo equipment, television receivers, etc. could be installed through the walls of the floor well. When the operator desired, he would maneuver a disguised hydraulic lever and lower the floor to the point desired. A hinged staircase could even be designed to automatically make its presence known as the floor was lowered. An alternate push of the hydraulic lever would bring the flooring section even again. I don't know really what the advantage would be in having a set-up such as this, but it sure would impress a lot of people. The spy in the movie seemed to be doing alright for himself. Seriously, this system could be made to work but a lot more thought would have to go into the finer details such as how to lock the floor in position should the hydraulic pressure suddenly fail, how to keep the floor section on track, etc. A lot of safety measures would certainly be required here. Getting the hydraulic system would be another problem, although some can be had very cheaply from some service stations about to go out of business. Hauling it and installing it would be the real problem. This would be a highly complex and costly system.

Another design which has been only partially tested and borders on the unusual is a little more down to earth. How many times have you awakened late on a sub-zero morning, rushed to get dressed, and then had to wait in a freezing automobile while the motor was warming up? With this system, you can start your car five or ten minutes before leaving the house without having to go outside and, manually, perform the starting procedure.

Depicted in Fig. 14-19, this system will have to be modified to conform to the starting switch lead arrangement and the priming characteristics of the various makes of automobiles. A good intruder alarm is also recommended if your car cannot be safely locked away-in a garage. This is remote control through radio tone activation using simple transmitting and receiving circuits. Many of these were discussed earlier in this text. This is a two channel system. One channel controls the priming solenoid which is located beneath the driver's seat. Before the driver leaves his car at night, he engages the control arm of this device with the accelerator pedal. The control arm is mounted to the solenoid armature by a hinged coupling which can be locked into place. This is shown in Fig. 14-20. The solenoid is a pushing type which forces the armature forward when activated. When the solenoid receives power from its relay which is switched on by solid-state components, it moves the control arm forward and holds it there for as long as the operator keeps the prime control closed on his transmitter. This continually transmits the tone which activates the solenoid switch.

After priming has been accomplished, the start control on the transmitter is depressed. This engages a heavy-duty relay which is a multicontact type (usually four to six contact functions). When the relay closes, the ignition system is activated and the starter motor is energized. If you can hear the engine from your location, you will know when to release the start control, but if you can't, you could destroy the internal starting solenoid mounted in the engine compartment. To overcome this problem, a miniature relay is

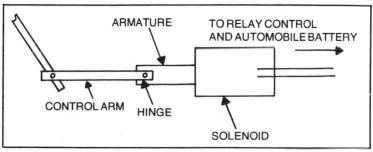

Fig. 14-20. Solenoid and control arm connections.

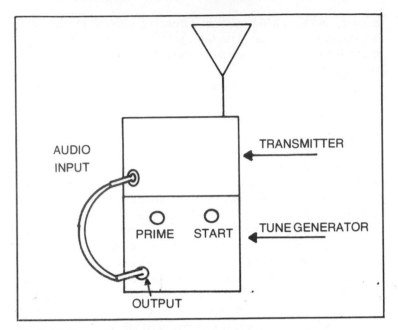

Fig. 14-21. Connection of tone generator to transmitter.

connected with its contacts in series with the DC power to the starting relay coil. This miniature relay has normally closed contacts and is a latching type. Its coil is placed in parallel with the output of the alternator. When the alternator begins charging, the coil of the relay becomes energized, and its contacts open. This breaks the circuit to the starting relay and prevents damage to the starter system.

The transmitter and receiver can be units of your own choice and may be built, or commercial units can be used. Choose frequencies that are not in your area and de-sensitize the receiver by keeping the antenna as short as possible. The transmitter is coupled to a tone generator as shown in Fig. 14-21. The tones chosen for this design were 2000 and 3000 hertz, respectively. The lower frequency controls the priming solenoid while the 3000 hertz frequency triggers the starting relay. The twin-T tone generator of an earlier chapter was used. The output is coupled to the input of the transmitter. While not shown, a matching transformer may be necessary here.

Figure 14-22 shows the filter and switching circuit. Very narrow bandwidth filters were designed from toroidal inductors

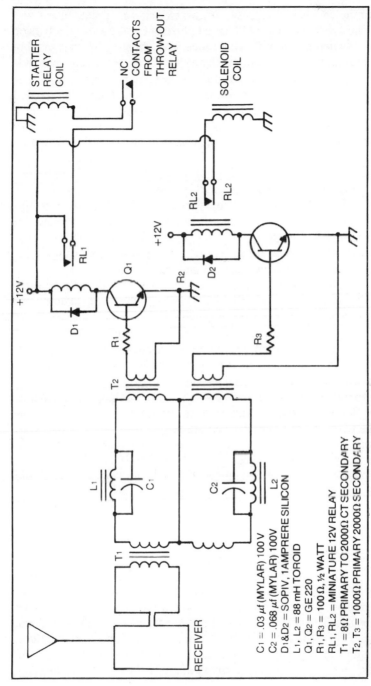

Fig. 14-22. Filter and switching circuits for remote-starting system.

C_1 = .03 µf (MYLAR) 100 V
C_2 = .068 µf (MYLAR) 100V
D_1 & D_2 = SOPIV, 1AMPRERE SILICON
L_1, L_2 = 88 mH TOROID
Q_1, Q_2 = GE 220
R_1, R_3 = 100 Ω, ½ WATT
RL_1, RL_2 = MINIATURE 12V RELAY
T_1 = 8Ω PRIMARY TO 2000Ω CT SECONDARY
T_2, T_3 = 1000Ω PRIMARY 2000Ω SECONDARY

paralleled by high-quality, mylar capacitors. These filters are critical, and the tones will have to be right on frequency to trigger the switching circuit. The tones are adjusted by the variable controls at the generator. The toroids are the common 88-millihenry varieties and should be easy to obtain from the different electronic parts catalogs. An output transformer is coupled between the receiver audio circuit and the filters. This is a component with a center-tapped secondary. Two other transformers couple the output from the filters to the transistor switches. These last transformers may be omitted, and the filter output connected directly to the bases of the transistors, but the indicated arrangement provides more exacting filtering.

When the base of the appropriate transistor is driven by the filter output, it conducts and the circuit is closed, activating the desired function. Once the automobile is started, the starter relay is disengaged from further control, but the solenoid can still be activated to idle the engine at a higher speed if desired.

This circuit will certainly require some modifications to work with different types of automobiles. Most of the changes, however, will be made at the starting relay which may require a different set of triggering functions. The circuit can be built on a fairly large piece of vector board and mounted in an aluminum box which should be grounded to the automobile chassis. A disconnect switch or switches can be placed in the appropriate positions in the automotive circuit, depending on the starter relay connections, to completely disengage the remote circuit from the automobile electrical system when normal starting of the car is desired. One note of caution: Since the automobile may be garaged and started in an enclosed area, be extremely cautious of the possibility of carbon monoxide poisoning. When the car is started, make certain the garage door has been opened beforehand. The automatic garage door circuit discussed earlier in this chapter can be used to good advantage here.

AUXILIARY WATER PUMP CONTROL

Some areas which are subject to torrential downpours create problems with basements flooding. Even though the traditional sump pump may be used in these situations, sometimes this is not enough. The sump may overflow, and the single pump just cannot handle the amount of water flowing in. This is especially bothersome and very costly if you have to be away from your home for long periods of time. An auxiliary pump can be used to aid the sump

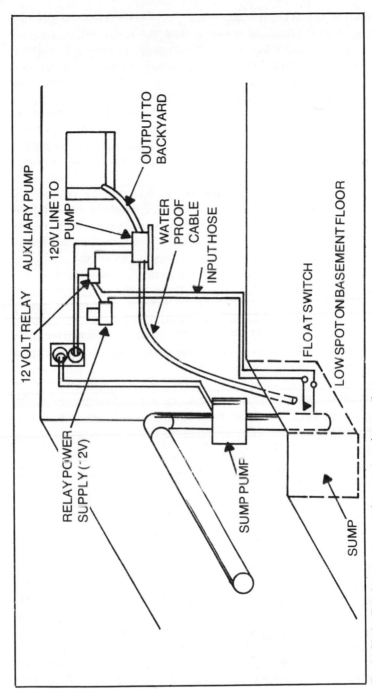

Fig. 14-23. Auxilliary basement pump placements and connections.

355

pump in removing flooding water. The auxiliary pump must be automatically triggered when the sump begins to overflow.

Figure 14-23 shows a practical circuit arrangement which is simple but effective. The auxiliary pump is mounted at a high spot in the basement, on the side of a wall if possible. It is connected to the AC wiring and is triggered by a miniature float switch located at a low part of the basement floor, near the sump. When the sump overflows, the float switch will be activated before the water becomes too deep. The auxiliary pump is activated and pumps the overflow water out of the basement and into the backyard through its output hose.

A 12 VDC controls the auxiliary pump motor and receives its power from a small DC supply acting off the AC line. The float switch controls the DC power supply output to the relay. In this manner, no 120-volt circuits are present on the basement floor which could prove hazardous if short circuited by water.

When the float switch is submerged, its contacts close. This supplies power to the 12-volt relay whose contacts also close. This completes the circuit from the AC line to the auxiliary pump motor, and the second pumping action begins. When the water has been mostly removed from the basement floor, the float switch opens its contacts again, and all power is removed from the 12-volt relay and, in turn, from the pump motor.

The original sump pump will do most of the general work as before, but when more water is coming in than it can pump out, the auxiliary pump aids the first one, and a faster removal is accomplished, possibly preventing heavy water damage.

SUMMARY

Literally millions of remote control applications are easily accomplished around the home. A system designed to perform one, specific function may be made to serve in many different remote control applications which are as varied as the varied number of work loads. This chapter has dealt with only a few of the many uses of remote control machines and electronics. Application is the key. The reader should have a good understanding of what remote control is, how it works, and how it can be applied. Imagination will play a large role now, coupled with the knowledge gained from this text. Abide by the laws of physics and almost anything can be accomplished. Abide by the laws of resourcefulness, and these accomplishments may cost little or nothing. Every function you attempt manually around your home can probably be done by a remote controlled machine and electronic circuit. Push your imagination and skill to the limits. If you fail, try again after studying the problem at length. Sooner or later, you will arrive at a workable solution to getting a job done with you as the remote controlling factor.

Appendix
Information For
Electronic Project Construction

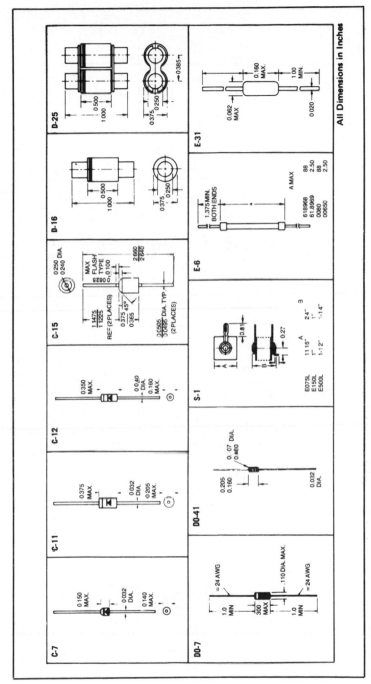

Fig. A-1. Diode case diagrams (courtesy of Workman).

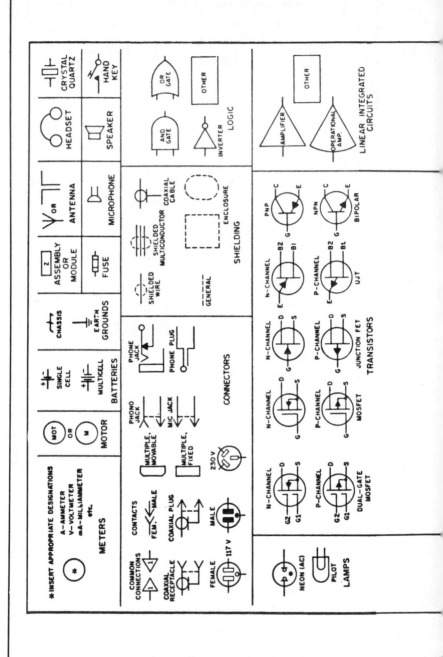

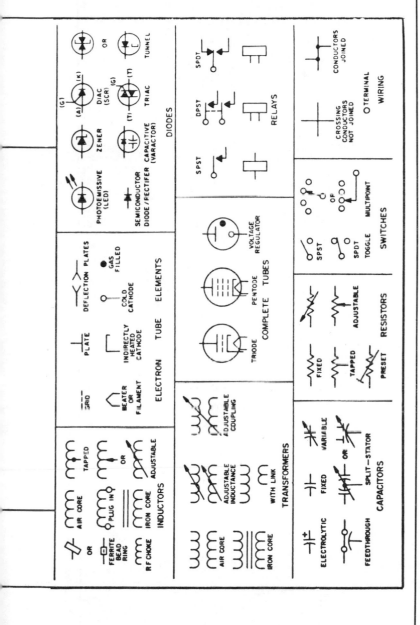

Fig. A-2. Commonly used electronic schematic diagrams (courtesy of American Radio Relay League Inc.).

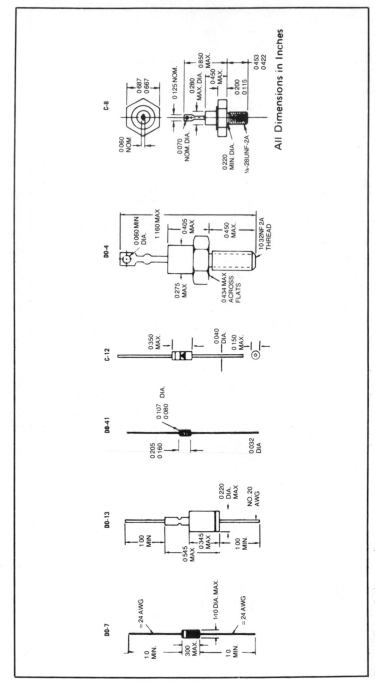

Fig. A-3. Zener diode case diagrams.

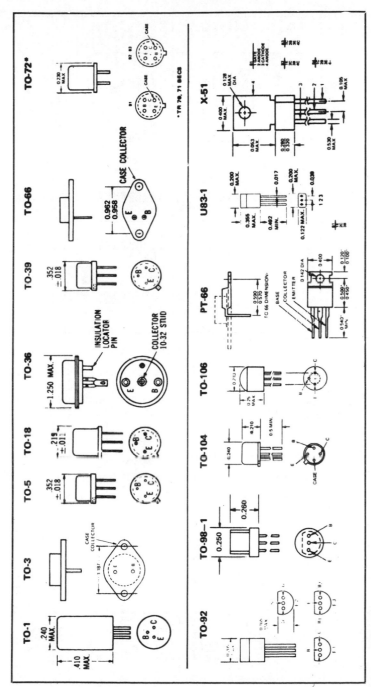

Fig. A-4. Transistor case diagrams.

363

Specification	Pushbutton	Toggle	Lighted Rocker
Contact Ratings	1A resistive load @ 115 VAC or 28 VDC or 0.5A	5 amps resistive load @ 115 VAC or 28 VDC, or 2 amps @ 250 VAC	5 amps resistive load @ 115 VAC or 28 VDC or 2 amps @ 250 VAC
Electrical Life	1,000,000 make-and-break cycles	1000,000 make-and-break cycles minimum on all models ending in 01 only, all other models 40,000 cycles	1000,000 cycles at rated loads
Initial Contact Resistance	10 milliohms max.	10 milliohms max. @ 2-4 VDC; 1 amp (for silver)	Under 5 milliohms @ 3 VDC, 100 milli-amp load
Insulation Resistance	10,000 megohms min.	1,000 megohms min.	20,000 megohms
Dielectric Strength	1,000 volts rms @ sea level	1,000 volts rms @ sea level	1,000 volts rms @ sea level
Case Materials	General purpose phenolic	General purpose phenolic	Diallyl Iso-Phthalate
Contact Material		Coin silver	Coin silver
Operating Lever		Bright, chrome-plated brass	
Mounting Bushing		Cadmium w/clear, chromate over brass	
Housing		Stainless steel	Stainless steel
Standard Hardware Supplied		2 hex mounting nuts; 1 internal tooth lockwasher; 1 locking ring	Nylon trim Bezel
Mounting Hole	1/4" dia.	1/4" dia.	.750" square

PUSHBUTTON SWITCHES

Cat. No.	Type	Contacts
PB532	SPST	Normally open
PB534	SPST	Normally closed

TOGGLE SWITCHES

Cat. No.	Type	Positions & Circuits			"A" Dimension
TS101	SPDT	ON	NONE	ON	0.270
TS103	SPDT	ON	OFF	ON	0.270
TS105	SPDT	(ON)	OFF	(ON)	0.270
TS201	DPDT	ON	NONE	ON	0.450
TS203	DPDT	ON	OFF	ON	0.450
TS211	DPDT	ON	ON	ON	0.450
TS301	3 PDT	ON	NONE	ON	0.650
TS401	4 PDT	ON	NONE	ON	0.850

(ON) – momentary

LIGHTED ROCKER SWITCHES

Cat. No.	Type	Positions & Circuits			Color
LRS101-19	SPDT	ON	NONE	ON	White
LRS101-39	SPDT	ON	NONE	ON	Red
LRS103-19	SPDT	ON	OFF	ON	White
LRS103-39	SPDT	ON	OFF	ON	Red
LRS201-19	DPDT	ON	NONE	ON	White
LRS201-39	DPDT	ON	NONE	ON	Red
LRS203-19	DPDT	ON	OFF	ON	White
LRS203-39	DPDT	ON	OFF	ON	Red

COLOR CAPS

Cat. No.	
TSC12	12 assorted vinyl color caps

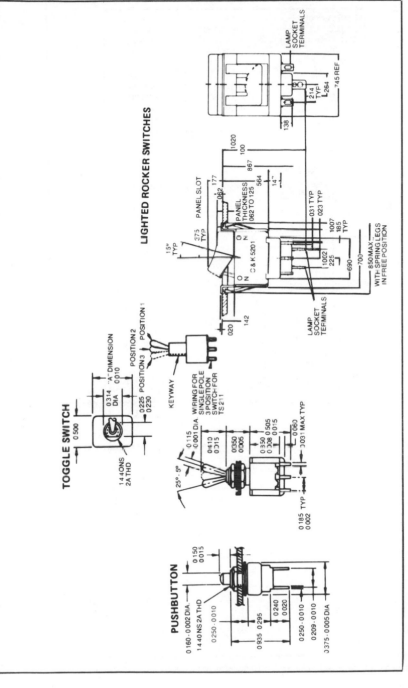

Fig. A-5. Subminiature switches (courtesy of Workman).

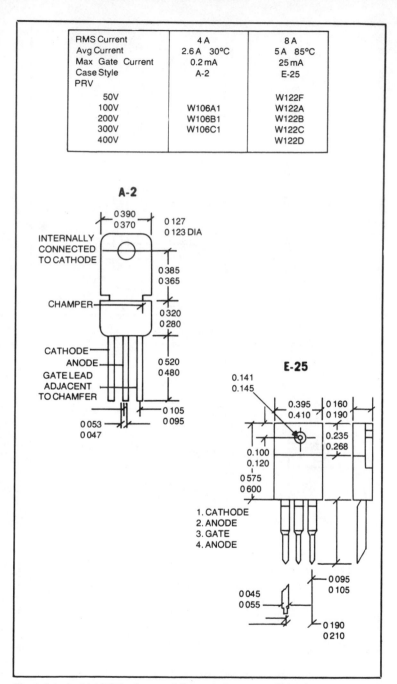

RMS Current	4 A	8 A
Avg Current	2.6 A 30°C	5 A 85°C
Max Gate Current	0.2 mA	25 mA
Case Style	A-2	E-25
PRV		
50V		W122F
100V	W106A1	W122A
200V	W106B1	W122B
300V	W106C1	W122C
400V		W122D

A-2

0 390
0 370

0 127
0 123 DIA

INTERNALLY
CONNECTED
TO CATHODE

0 385
0 365

CHAMPER

0 320
0 280

CATHODE

ANODE

GATE LEAD
ADJACENT
TO CHAMFER

0 520
0 480

0 105
0 095

0 053
0 047

E-25

0.141
0.145

0.395
0.410

0 160
0 190

0.235
0.268

0.100
0.120

0 575
0 600

1. CATHODE
2. ANODE
3. GATE
4. ANODE

0 095
0 105

0 045
0 055

0 190
0 210

Fig. A-6. Industrial SCR ratings and diagrams (courtesy of Workman).

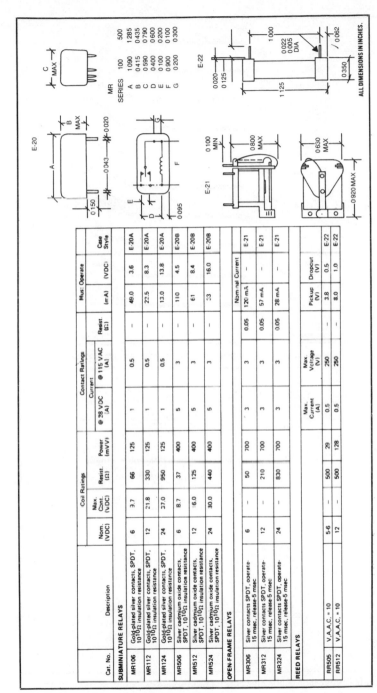

Fig. A-7. Circuit board relays (courtesy of Workman).

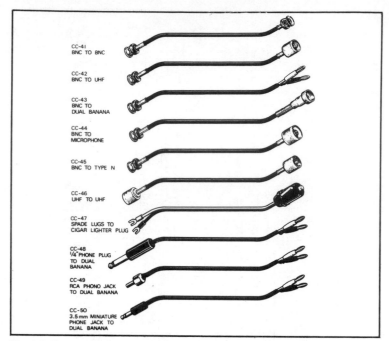

Fig. A-8. Types of cables and connectors used in radio-control applications.

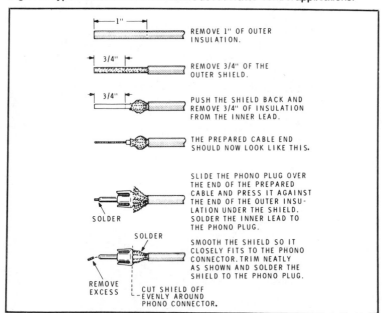

Fig. A-9. Preparation of coaxial cable for phono plug.

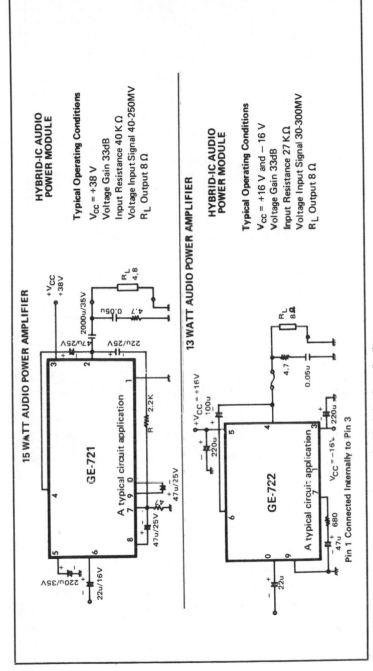

Fig. A-10. Hybrid IC audio power modules and typical circuit applications.

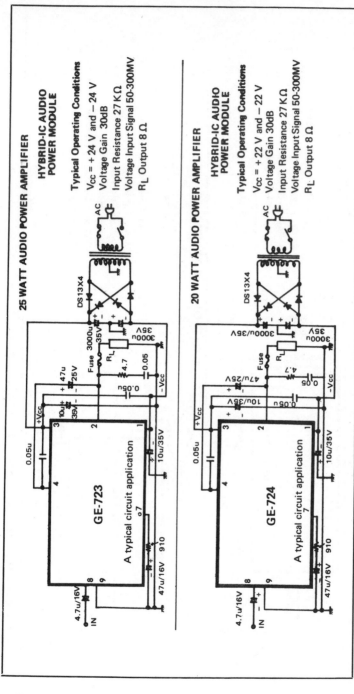

Fig. A-11. Hybrid IC audio power modules and typical circuit applications.

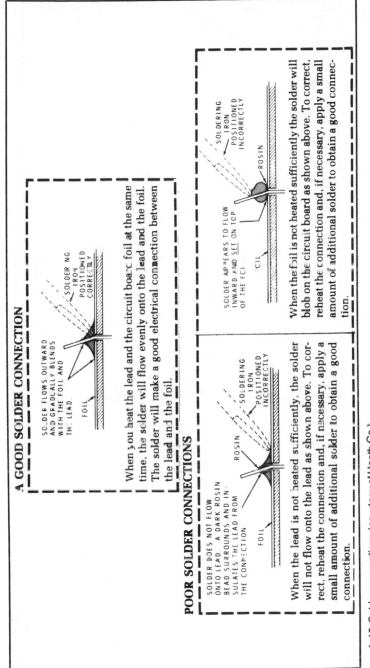

A GOOD SOLDER CONNECTION

SOLDER FLOWS OUTWARD
AND GRADUALLY BLENDS
WITH THE FOIL AND
THE LEAD.

SOLDERING
IRON
POSITIONED
CORRECTLY

FOIL

When you heat the lead and the circuit board foil at the same time, the solder will flow evenly onto the lead and the foil. The solder will make a good electrical connection between the lead and the foil.

POOR SOLDER CONNECTIONS

SOLDER DOES NOT FLOW
ONTO LEAD. A DARK ROSIN
BEAD SURROUNDS AND IN-
SULATES THE LEAD FROM
THE CONNECTION.

ROSIN

SOLDERING
IRON
POSITIONED
INCORRECTLY

FOIL

When the lead is not heated sufficiently, the solder will not flow onto the lead as shown above. To correct, reheat the connection and, if necessary, apply a small amount of additional solder to obtain a good connection.

SOLDER APPEARS TO FLOW
INWARD AND SET ON TOP
OF THE FOIL.

SOLDERING
IRON
POSITIONED
INCORRECTLY

ROSIN

FOIL

When the foil is not heated sufficiently the solder will blob on the circuit board as shown above. To correct, reheat the connection and, if necessary, apply a small amount of additional solder to obtain a good connection.

Fig. A-12. Solder connections (courtesy of Heath Co.).

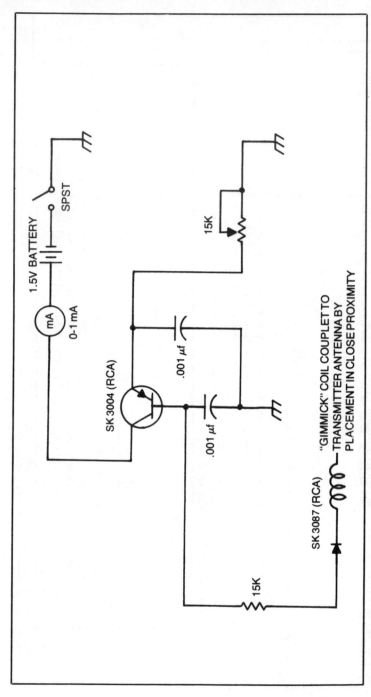

Fig. A-13. Relative-strength meter for low-power transmitters.

Table A-1. Resistor Color Code.

Color	Significant Figure	Decimal Multiplier
Black	0	1
Brown	1	10
Red	2	100
Orange	3	1,000
Yellow	4	10,000
Green	5	100,000
Blue	6	1,000,000
Violet	7	10,000,000
Gray	8	100,000,000
White	9	1,000,000,000
Gold	—	0.1
Silver	—	0.01

Table A-2. Relative Conductance of Various Metals.

Substance	Relative conductance (Silver = 100%)
Silver	100
Copper	98
Gold	78
Aluminum	61
Tungsten	32
Zinc	30
Platinum	17
Iron	16
Lead	15
Tin	9
Nickel	7
Mercury	1
Carbon	0.05

Table A-3. Resistivities of Conductors at 0° C.

Resistivities of Conductors at 0° C			
	Microhms		Round Wires
Substance	Centimeter cube	Inch cube	Ohms-Circular mils per foot
Aluminum	3.21	1.26	19.3
Carbon	4000 to 10,000	1600 to 2800	24,00 to 42,000
Constantan (Cu 60%, Ni 40%)	49	19.3	295
Copper	1.72	0.68	10.4
Iron	12 to 14	4.7 to 5.5	72 to 84
Lead	20.8	8.2	125
Manganin (Cu 84%, Ni 4%, Mn 12%)	43	16.9	258
Mercury	95.76	37.6	575
Nichrome (Ni 60%, Cr 12%, Fe 26%, Mn 2%)	110	43	660
Platinum	11.0	4.3	66
Silver	1.65	0.65	9.9
Tungsten	5.5	2.15	33
Zinc	6.1	2.4	36.7

Table A-4. Horsepower and Electrical Power Equivalents.

Unit	Equivalents
1 H.P	745.7 Watts
1 H.P	0.746 Kw.
1 H.P	33,000 Ft.-Lbs. per Min.
1 H.P	550 Ft.-Lbs. per Sec.
1 H.P	2,545 B.T.U per Hr.
1 H.P	0.175 Lbs. Carbon oxidized per Hr.
1 H.P	17 Lbs. Water per Hr. heated from 62-212° F.
1 H.P	2.64 Lbs. Water per Hr. evaporated from and at 212°F.
1 Kw	1,000 Joules per Sec.
1 Kw	1.34 H.P.
1 Kw	44,250 Ft. Lbs. per Min.
1 Kw	737.3 Ft.-Lbs. per Sec.
1 Kw	3,412 B.T.U. per Hr.
1 Kw	0.227 Lbs. Carbon oxidized per Hr.
1 Kw	22.75 Lbs. Water per Hr. heated from 62-212°F.
1 Kw	3.53 Lbs. Water per Hr. evaporated from and at 212°F.

374

Table A-5. Horsepower Versus Watts.

HP	Watts	HP	Watts
.01	7.457	.26	193.882
.02	14.914	.27	201.339
.03	22.371	.28	208.796
.04	29.828	.29	216.253
.05	37.285	.30	223.710
.06	44.742	.31	231.167
.07	52.199	.32	238.624
.08	59.656	.33	246.081
.09	67.113	.34	253.538
.10	74.570	.35	260.995
.11	82.027	.36	268.452
.12	89.484	.37	275.909
.13	96.941	.38	283.366
.14	104.398	.39	290.823
.15	111.855	.40	298.280
.16	119.312	.41	305.737
.17	126.769	.42	313.194
.18	134.226	.43	320.651
.19	141.683	.44	328.108
.20	149.140	.45	335.565
.21	156.597	.46	343.022
.22	164.054	.47	350.479
.23	171.511	.48	357.936
.24	178.968	.49	365.393
.25	186.425	.50	372.850
.51	380.307	.76	556.732
.52	387.764	.77	574.189
.53	395.221	.78	581.646
.54	402.678	.79	589.103
.55	410.135	.80	596.560
.56	417.592	.81	604.017
.57	425.049	.82	611.474
.58	432.506	.83	618.931
.59	439.963	.84	626.388
.60	447.420	.85	633.845
.61	454.877	.86	641.302
.62	462.334	.87	648.759
.63	469.791	.88	656.216
.64	477.248	.89	663.673
.65	484.705	.90	671.130
.66	492.162	.91	678.587
.67	499.619	.92	686.044
.68	507.076	.93	693.501
.69	514.533	.94	700.958
.70	521.990	.95	708.415
.71	529.447	.96	715.872
.72	536.904	.97	723.329
.73	544.361	.98	730.786
.74	551.818	.99	738.243
.75	559.275	1.00	745.700

Table A-6. Watts Versus Horsepower.

Watts	HP	Watts	HP	Watts	HP
1	.001341	34	.045594	67	.089847
2	.002682	35	.046935	68	.091188
3	.004023	36	.048278	69	.092529
4	.005364	37	.049617	70	.093870
5	.006705	38	.050958	71	.095211
6	.008046	39	.052299	72	.096552
7	.009387	40	.053640	73	.097893
8	.010728	41	.054981	74	.099234
9	.012069	42	.056322	75	.100575
10	.013410	43	.057663	76	.101916
11	.014751	44	.059004	77	.103257
12	.016092	45	.060345	78	.104598
13	.017433	46	.061686	79	.105939
14	.018774	47	.063027	80	.107280
15	.020115	48	.064368	81	.108261
16	.021456	49	.065709	82	.109962
17	.022797	50	.067050	83	.111303
18	.024138	51	.068391	84	.112644
19	.025479	52	.069732	85	.113985
20	.026820	53	.071073	86	.115326
21	.028161	54	.072414	87	.116667
22	.029502	55	.073755	88	.118008
23	.030843	56	.075096	89	.119349
24	.032184	57	.076437	90	.120690
25	.033525	58	.077778	91	.122031
26	.034866	59	.079119	92	.123372
27	.036207	60	.080460	93	.124713
28	.037548	61	.081801	94	.126054
29	.038889	62	.083142	95	.127395
30	.040230	63	.084483	96	.128736
31	.041571	64	.085824	97	.130077
32	.042912	65	.087165	98	.131418
33	.044253	66	.088506	99	.132759
				100	.134100

Table A-7. Light-Activated and Light-Controlled Devices.

Cat. No.	Description	Output* Volts	Output* Current
SILICON SOLAR CELLS			
S1M	Molded cases 1-⅛ × 1-⅛ × 3/16	0.3-0.4	10-16 mA
S3M	Molded case 1-⅛ × 1-⅛ × 3/16	0.6-0.85	10-16 mA
S4M	Molded case 1-⅛ × 1-⅛ × 3/16	0.3-0.4	25-40 mA
S6M	Molded case 1-⅛ × 1-⅛ × 3/16	1.5	5-8 mA
S7M	Molded case 1-⅛ × 1-⅛ × 3/16	3.0	5-8 mA

*Typical mW/cm^2, M = 1

Cat. No.	Description	Output* Volts	Output* Current
SELENIUM PHOTO CELLS			
WB2M	Cell on mounting bracket	0.2-0.4	77 μA
WB3M	Molded case 1-⅛ × 1-⅛ ×3/16	0.2-0.4	77 μA

*Typical   100 fc, 100 ohms

CADMIUM SULPHIDE PHOTOCONDUCTIVE CELLS

Cat. No.	Description	Max. Volts.	Max. Resistance Dark	Max. Resistance 100 fc
CS120	Molded case 1-⅛ × 1-⅛ × 3/16	20V	1,600Ω	0.11Ω

Table A-8. Common Motor Troubles and Repairs.

Motor Fails To Start	
Cause	**Remedy**
Fuses blown, switch open, broken or poor connections, or no voltage on line.	Check for proper voltage at motor terminals. Examine fuses, switches, and connections between motor terminals and points of service. Look for broken wires, bad connections, corroded fuse holders. Repair or replace as necessary.
Defective motor windings.	Locate and repair.[1]

Motor Hums But Will Not Start	
Starting winding switch does not close.	Clean or replace and lubricate if needed.
Defective starting capacitor.	Replace.[1]
Open rotor or stator coil.	Locate and repair.[1]
Motor overloaded.	Lighten load. Check for low voltage.
Overloaded line or low voltage.	Reduce electrical load. Check wiring. Increase wire size. Notify power company.
Bearings worn so that rotor rubs on starter.	Replace bearings. Center rotor in stator bore.[1]
Bearings too tight or lack of proper lubrication.	Clean and lubricate bearings. Check end bells for alignment.
Burned or broken connections.	Locate and repair.

Motor Will Not Start With Rotor In Certain Position	
Burned or broken connections; open rotor or stator coil.	Inspect, test, and repair.[1]

Motor Runs But Then Stops	
Motor overloaded.	Lighten motor load. Check for low voltage.
Defective overload protection.	Locate and replace.[1]

Slow Acceleration	
Overloaded motor.	Lighten motor load.
Poor connections.	Test and repair.
Low voltage or overloaded line.	Lighten line load. Increase size of line wire.[1]
Defective capacitor.	Replace.[1]

Excessive Heating	
Cause	**Remedy**
Overloaded motor.	Reduce motor load.
Poor or damaged insulation; broken connections; or grounds or short circuits.	Locate and repair.[1]
Wrong connections.	Check wiring diagram of motor.
Worn bearings or rotor rubs on stator.	Renew or repair bearings. Check end bell alignment.[1]

Table A-8. Common Motor Troubles and Repair (continued from page 378).

Excessive Heating	
Cause	**Remedy**
Bearings too tight or lack of proper lubrication.	Clean and lubricate bearings. Check end bell alignment.
Belt too tight.	Slacken belt.
Motor dirty or improperly ventilated.	Clean motor air passages.
Defective capacitor.	Replace.
Excessive Vibration	
Unbalanced rotor or load.	Rebalance rotor or load.
Worn bearings.	Replace.[1]
Motor misaligned with load.	Align motor shaft with load shaft.
Loose mounting bolts.	Tighten.
Unbalanced pulley.	Have pulley balanced or replaced.
Uneven weight of belt.	Get new belt.
Low Speed	
Overloaded.	Reduce load.
Wrong or bad connections.	Check for proper voltage connections and repair.
Low voltage, overloaded line, or wiring too small.	Reduce load. Increase size of wire.[1]

[1] These repairs should be made by an experienced electrician.

Table A-9. Special Electronic Components.

Cat. No.	Max. Current	Max. Voltage	Case Style
HOBBY PROJECT SCR's			
SCR03 SCR04	9 9	50 200	E-64 E-64
PLASTIC SCRs			
W106Y1 W106A1 W106B1 W106C1 W122F W122A W122B W122C	2 2 2 2 5 5 5 5	30 100 200 300 50 100 200 300	A-2 A-2 A-2 A-2 E-25 E-25 E-25 E-25
TRIAC			
IRT82	8	200	E-25
DIAC			
IRD54	500 mW	32	C-10
UNIJUNCTION TRANSISTOR			
IR2160	I_e 50 mA	30	TO-92 (E3)

Table A-10. Recommended Wire Sizes for Varying Circuit Lengths and Current.

Length (ft)	30	40	50	60	70	80	90	100	125	150	175	200	225	250	275	300	350	400
Amper Load							Wire Sizes											
5	14	14	14	14	14	14	12	12	12	10	10	10	8	8	8	8	6	6
6	14	14	14	14	14	12	12	12	10	10	10	6	8	8	8	6	6	6
7	14	14	14	14	12	12	12	12	10	10	8	8	8	8	6	6	6	6
8	13	13	13	12	12	12	12	10	10	8	8	8	8	6	6	6	6	4
9	14	14	12	12	12	12	10	10	10	8	8	8	6	6	6	4	6	4
10	14	14	12	12	12	10	10	10	8	8	8	6	6	6	6	4	4	4
12	14	12	12	10	10	10	8	8	8	6	6	6	6	4	4	4	4	2
14	12	12	12	10	10	8	8	8	6	6	6	6	4	4	4	4	2	2
16	12	12	10	10	10	8	8	8	6	6	6	4	4	4	4	2	2	2
18	12	12	10	10	8	8	8	6	6	6	4	4	4	4	2	2	2	2
20	12	10	10	8	8	6	8	6	6	4	4	4	2	2	2	2	2	1
25	10	10	8	8	8	6	6	6	4	4	4	2	2	2	1	1	1	0
30	10	8	8	8	6	6	6	4	4	2	2	2	2	1	0	0	0	00
35	10	8	8	6	6	6	4	4	2	2	2	2	1	1	0	0	00	00
40	8	8	6	6	6	4	4	4	2	2	2	1	1	0	0	00	00	000
45	8	8	6	6	4	4	4	4	2	2	1	1	0	0	00	00	000	000
50	8	6	6	4	4	2	4	2	1	1	1	0	0	00	00	000	000	0000
60	6	6	4	4	4	2	2	2	1	1	0	00	00	000	000	000	250	250
70	6	6	4	3	2	3	2	2	1	0	00	00	000	000	300	300	00	300
80	6	4	4	2	3	1	2	1	0	00	000	0000	0000	250	300	350	000	000
90	6	4	4	1	2	0	1	1	00	000	0000	250	250	250	300	400	0000	250
100	4	4	2	0	1	00	1	0	000	0000	300	300	300	300	350	500	250	300
130	4	2	2		0	000	00	00	0000	250	350	400	350	350	500		300	350
160	4	2	1		0		00	000	250	300	500	500	400	500	500			
200	2	1	0				000	0000		400	500	500	500	500				

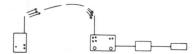

Index

Edited by Raymond A. Collins